Lecture Notes in Networks and Systems   895

Series Editor

Janusz Kacprzyk ⓘ, *Systems Research Institute, Polish Academy of Sciences, Warsaw, Poland*

Advisory Editors

Fernando Gomide, *Department of Computer Engineering and Automation—DCA, School of Electrical and Computer Engineering—FEEC, University of Campinas—UNICAMP, São Paulo, Brazil*
Okyay Kaynak, *Department of Electrical and Electronic Engineering, Bogazici University, Istanbul, Türkiye*
Derong Liu, *Department of Electrical and Computer Engineering, University of Illinois at Chicago, Chicago, USA*
   Institute of Automation, Chinese Academy of Sciences, Beijing, China
Witold Pedrycz, *Department of Electrical and Computer Engineering, University of Alberta, Alberta, Canada*
   Systems Research Institute, Polish Academy of Sciences, Warsaw, Poland
Marios M. Polycarpou, *Department of Electrical and Computer Engineering, KIOS Research Center for Intelligent Systems and Networks, University of Cyprus, Nicosia, Cyprus*
Imre J. Rudas, *Óbuda University, Budapest, Hungary*
Jun Wang, *Department of Computer Science, City University of Hong Kong, Kowloon, Hong Kong*

The series "Lecture Notes in Networks and Systems" publishes the latest developments in Networks and Systems—quickly, informally and with high quality. Original research reported in proceedings and post-proceedings represents the core of LNNS.

Volumes published in LNNS embrace all aspects and subfields of, as well as new challenges in, Networks and Systems.

The series contains proceedings and edited volumes in systems and networks, spanning the areas of Cyber-Physical Systems, Autonomous Systems, Sensor Networks, Control Systems, Energy Systems, Automotive Systems, Biological Systems, Vehicular Networking and Connected Vehicles, Aerospace Systems, Automation, Manufacturing, Smart Grids, Nonlinear Systems, Power Systems, Robotics, Social Systems, Economic Systems and other. Of particular value to both the contributors and the readership are the short publication timeframe and the world-wide distribution and exposure which enable both a wide and rapid dissemination of research output.

The series covers the theory, applications, and perspectives on the state of the art and future developments relevant to systems and networks, decision making, control, complex processes and related areas, as embedded in the fields of interdisciplinary and applied sciences, engineering, computer science, physics, economics, social, and life sciences, as well as the paradigms and methodologies behind them.

Indexed by SCOPUS, INSPEC, WTI Frankfurt eG, zbMATH, SCImago.

All books published in the series are submitted for consideration in Web of Science.

For proposals from Asia please contact Aninda Bose (aninda.bose@springer.com).

Mostafa Al-Emran · Jaber H. Ali · Marco Valeri ·
Alhamzah Alnoor · Zaid Alaa Hussien
Editors

Beyond Reality: Navigating the Power of Metaverse and Its Applications

Proceedings of 3rd International
Multi-Disciplinary Conference - Theme:
Integrated Sciences and Technologies
(IMDC-IST 2024) Volume 1

 Springer

Editors
Mostafa Al-Emran
The British University in Dubai
Dubai, United Arab Emirates

Jaber H. Ali
Southern Technical University
Basrah, Iraq

Marco Valeri
Faculty of Economics
University Niccolò Cusano
Rome, Italy

Alhamzah Alnoor
Southern Technical University
Basrah, Iraq

Zaid Alaa Hussien
Southern Technical University
Basrah, Iraq

ISSN 2367-3370   ISSN 2367-3389 (electronic)
Lecture Notes in Networks and Systems
ISBN 978-3-031-51715-0   ISBN 978-3-031-51716-7 (eBook)
https://doi.org/10.1007/978-3-031-51716-7

This Springer imprint is published by the registered company Springer Nature Switzerland AG
The registered company address is: Gewerbestrasse 11, 6330 Cham, Switzerland

Paper in this product is recyclable.

Preface

The Metaverse refers to a three-dimensional virtual space involving social connections. Its management applications landscape is undergoing significant changes not only as a result of global restrictions related to COVID-19 but also due to disruptive innovations in the digital realm. Metaverse has changed people's perception of virtual experiences for management applications. Many practitioners focus on creating virtual experiences and products for consumers. The 3rd International Multi-Disciplinary Conference Theme: Integrated Sciences and Technologies (IMDC-IST 2024) is held to address the theme "Beyond reality: Navigating the power of Metaverse and its applications". The IMDC-IST 2024 brings together a wide range of researchers from different disciplines. It seeks to promote, encourage, and recognize excellence in scientific research related to Metaverse and virtual reality applications. The main aim of the IMDC-IST 2024 is to provide a forum for academics, researchers, and developers from academia and industry to share and exchange their latest research contributions and identify practical implications of emerging technologies to advance the wheel of these solutions for global impact. In line with the Fourth Industrial Revolution goals and its impact on sustainable development, IMDC-IST 2024 is devoted to increasing the understanding and impact of the Metaverse on individuals, organizations, and societies and how Metaverse applications have recently reshaped these entities.

The IMDC-IST 2024 attracted 65 submissions from different countries worldwide. Out of the 65 submissions, we accepted 29, representing an acceptance rate of 44.61%. About 15 papers are accepted in Volume 1. The chapters of Volume 1 collectively focus on a multifaceted exploration of the Metaverse and its intersection with various sectors and technologies. They examine the evaluation of Metaverse tools through a privacy-centric lens using decision-making models, delve into the current research trends within the Metaverse, and discuss the enhancement of human-virtual interactions via virtual reality. Volume 1 also explores the creative aspects of virtual reality in education, particularly in language learning, and how Metaverse applications can be integrated into educational settings to improve learning experiences. Further, it investigates user behavior within the Metaverse, considering different factors and technology acceptance models, and discusses the social sustainability of the Metaverse through user satisfaction. There is a forward-looking analysis of the potential impacts of the Metaverse on higher education quality and an interesting synthesis of how characteristics of AI, exemplified by ChatGPT, can align with Metaverse attributes like knowledge sharing and ethical considerations.

The book also ventures into more niche territories, such as applying financial measures within the context of Industry 5.0 and the role of big data in enhancing supply chain management within the Metaverse framework. It touches upon integrating artificial intelligence and Metaverse techniques in financial practices and addresses the implications of digital assets accounting within this new virtual domain. Lastly, it presents insights into the influence of digital business strategies on customer experiences and investigates financial readiness concerning virtual technologies. Each chapter offers a unique perspective on the intricate relationship between the Metaverse and various domains, emphasizing practical applications and future directions.

Each submission is reviewed by at least two reviewers, who are considered experts in the related submitted paper. The evaluation criteria include several issues: correctness, originality, technical strength, significance, presentation quality, interest, and relevance to the conference scope. The conference proceedings are published in *Lecture Notes in Networks and Systems Series* by Springer, which has a high SJR impact. We acknowledge all those who contributed to the success of IMDC-IST 2024. We would also like to thank the reviewers for their valuable feedback and suggestions. Without them, it was impossible to maintain the high quality and success of IMDC-IST 2024.

Mostafa Al-Emran
Jaber H. Ali
Marco Valeri
Alhamzah Alnoor
Zaid Alaa Hussien

Organization

Conference General Chairs

Mostafa Al-Emran   The British University in Dubai, Dubai, UAE
Alhamzah Alnoor   Southern Technical University, Basrah, Iraq

Honorary Conference Chair

Rabee Hashem Al-Abbasi   Chancellor of Southern Technical University,
             Basrah, Iraq

Conference Organizing Chair

Jaber H. Ali   Dean of Management Technical College,
             Southern Technical University, Basrah, Iraq

Program Committee Chair

Marco Valeri   Faculty of Economics, Niccolo' Cusano
             University in Rome, Italy

Publication Committee Chairs

Mostafa Al-Emran   The British University in Dubai, Dubai, UAE
Jaber H. Ali   Dean of Management Technical College,
             Southern Technical University, Basrah, Iraq
Marco Valeri   Niccolo' Cusano University, Italy
Alhamzah Alnoor   Southern Technical University, Basrah, Iraq
Zaid Alaa Hussein   Southern Technical University, Basrah, Iraq

Conference Tracks Chairs

Sammar Abbas	Institute of Business Studies, Kohat University of Science and Technology, Pakistan
Gül Erkol Bayram	School of Tourism and Hospitality Management Department of Tour Guiding, Sinop University, Sinop, Turkey
Marcos Ferasso	Economics and Business Sciences Department, Universidade Autónoma de Lisboa, 1169-023 Lisboa, Portugal
Hussam Al Halbusi	Department of Management at Ahmed Bin Mohammad Military College, Doha, Qatar
Khai Wah Khaw	School of Management, Universiti Sains Malaysia, 11800, Pulau Pinang, Malaysia
Gadaf Rexhepi	Southeast European University, Tetovo, The Republic of Macedonia

Members of Scientific Committee

Hashem Nayef Hashem	Sothern Technical University, Technical College of Management, Basra, Iraq
Gul Erkol Bayram	Sinop University, School of Tourism and Hospitality Management Department of Tour Guiding, Turkey
Akram Mohsen Al-Yasiri	University of Karbala, Iraq
Hussan Al Halbusi	Department of Management, Ahmed Bin Mohammed Military College, Qatar
Muslim Allawi Al-Shibli	Maqal University, Iraq
Sammar Abbas	Institute of Business Studies, Kohat University of Science and Technology, Pakistan
Taher Mohsen Mansour	Shatt Al-Arab University, Iraq
Gadaf Rexhepi	Southern European University, The Republic of Macedonia
Abdul Hussein Tawfiq Shalabi	College of Administration and Economics, University of Basra, Iraq
Marcos Ferasso	Economic and Business Sciences Department, Universidade Autonoma de Lisboa, Portugal
Safaa Muhammad Hadi	Sothern Technical University, Technical College of Management, Basra, Iraq
Nabil Jaafar Al-Marsoumi	Al-Maqal University, Iraq
Muhammad Helou Daoud	Ministry of Higher Education, Iraq

Khai Wah Khaw	School of Management, Universiti Sains Malaysia, Malaysia
Hadi Abdel Wahab	College of Administration and Economics, University of Basra, Iraq
Salma Abdel Baqi	College of Information Technology, University of Basra, Iraq
Imad Abdel Sattar Salman	Southern Technical University, Basrah, Iraq

Publicity and Public Relations Committee

Zaid Alaa Hussein	Southern Technical University, Basrah, Iraq
Alhamzah Alnoor	Southern Technical University, Basrah, Iraq

Finance Chair

Jaber H. Ali	Dean of Management Technical College, Southern Technical University, Basrah, Iraq

Contents

Evaluation of Metaverse Tools Based on Privacy Model Using Fuzzy MCDM Approach

Nor Azura Husin[1]([✉]), Ali A. Abdulsaeed[2], Yousif Raad Muhsen[1,2],
Ali Shakir Zaidan[3], Alhamzah Alnoor[4], and Zahraa Raad Al-mawla[5]

[1] Department of Computer Science, Faculty of Computer Science and Information Technology,
Universiti Putra Malaysia, Serdang, Malaysia
n_azura@upm.edu.my
[2] Civil Department, College of Engineering, Wasit University, Kut, Iraq
{amuniem,yousif}@uowasit.edu.iq
[3] School of Management, Universiti Sains Malaysia, 11800 Pulau Pinang, Malaysia
sha3883@student.usm.my
[4] Management Technical College, Southern Technical University, Basrah, Iraq
Alhamzah.malik@stu.edu.iq
[5] Department of Education, University of Al-Mustansiriyah, 10001 Baghdad, Iraq
zahraa.raad@uomustansiriyah.edu.iq

Abstract. The issue of uncertainty in decision modelling for control engineering
tools that support industrial cyber-physical metaverse smart manufacturing systems (ICPMSMSs) can be classified as a multi-criteria decision-making challenge
for three primary reasons. Firstly, there are multiple components in the industry
cyber-physical system (ICPS) that need to be taken into account. Secondly, these
components vary in terms of their importance. Lastly, there is a variation in the
available data. Despite substantial efforts, neither of the control engineering tools
developed for ICPMSMSs has been capable of completely satisfying all ICPS
aspects. As a result, choosing the best possible tool to support ICPMSMSs is a
difficult process. Even though studies have evaluated control engineering tools,
there are still unresolved issues regarding information ambiguity and uncertainty.
In order to overcome these issues, this research improves the fuzzy weighted with
zero inconsistency (FWZIC) method by adding Hexagonal Fuzzy Numbers to the
process of assigning weights to the ICPS's multiple components. Subsequently, the
established Hexagonal-FWZIC method will be combined with the additive ratio
assessments (ARAS) method to handle the rank in modelling control engineering
tools. According to the results of the Hexagonal-FWZIC analysis, cybersecurity
is the most influential factor, while the digital twin component has the least. The
ARAS findings indicate that, of the ten tools studied, tools_1, tool_3, and tool_4
are the best three tools.

Keywords: FWZIC · ARAS · ICPMSMSs · Metaverse · Hexagonal-FWZIC

M. Al-Emran et al. (Eds.): IMDC-IST 2024, LNNS 895, pp. 1–20, 2023.
https://doi.org/10.1007/978-3-031-51716-7_1

1 Introduction

The phrase metaverse is a comparatively new supplement to the everyday common sense of technology academics and scientists likewise, it was originally employed in a Neal Stephenson novel named Snow Crash in 1992. The term meta originates from the Greek language and denotes a state of being that surpasses or transcends the current level of understanding or experience. It implies the existence of additional layers of knowledge or perception that can be further developed and explored (Hollensen et al., 2023; Atiyah et al., 2022; Gatea and Marina, 2016). The term metaverse pertains to a virtual environment that serves as a replacement for the physical world. The platform provides a comprehensive virtual reality (VR) encounter that encompasses prominent players, entities, interfaces, and social networks. The definition in question is corroborated by the scholarly contributions of Dionisio et al. (2013) and Smart et al. (2007). Dionisio et al. (2013) posit that the metaverse exhibits a number of fundamental features, such as pervasive presence, immersive verisimilitude, and capacity for expansion with respect to player count, intricate environments, and a wide array of interactive possibilities. The metaverse has emerged as a significant avenue for brand marketing, given the increased virtual engagement of consumers during the pandemic. It is anticipated that in the aftermath of the pandemic, the younger demographic will exhibit a heightened interest in digital ownership. According to Liederman (2021), the way individuals participate in the Metaverse realm is of equal significance to their involvement in the physical world.

The optimal realization of the metaverse is contingent upon the availability of high-speed internet connectivity. An illustrative example is the virtual reality platform "Second Life," which was introduced prior to the widespread adoption of smartphones and experienced a decline in popularity due, in part, to its inability to facilitate immediate, mobile interactions. Virtual reality and augmented reality are technologies that facilitate complex interactions with people, digital objects, and environments. The metaverse is presented in the novel as a virtual reality, a universe that uses the internet and augmented reality through software agents and avatars (Zhao et al., 2022; Hamid et al., 2021). Metaverse can be defined as a network of socially connected and persistent immersive platforms that enable seamless communication and dynamic interactions with digital objects. It represents a virtual environment that interfaces with the physical world through diverse applications and a three-dimensional (3D) representation. The metaverse incorporates avatars enabled by smart devices like smartphones, tablets, smart glasses, headsets, and other advanced technologies such as extended reality, digital twin, and artificial intelligence. These technologies facilitate people to connect, interact, and engage in a wide range of activities and tasks within the metaverse (Dwivedi et al., 2023; AL-Fatlawey et al., 2021). Its early version was a network of virtual worlds with teleportation between them. The contemporary metaverse is made up of social virtual reality systems that are immersive and work with open game environments, augmented reality settings that encourage collaboration, and massively multiplayer online games. This recognition provides a framework for understanding the potential socioeconomic implications of a fully realized, cross-platform metaverse. Hence, the metaverse is currently a hot topic in a variety of sectors, including business and management, information systems, social sciences, logistics, supply chain, and innovation, R&D among others.

Although information and communication technology have advanced at an unparalleled rate, metaverse applications in sectors connected to operations and supply chain management are still in the early stages (Dwivedi et al., 2023; Atiyah, 2022). However, in 2021 the term metaverse has obtained a number of attracts and advantages due to the amazing huge of investment and level of satisfaction viewed by various national firms in 2021 on the metaverse related to the megaproject. As a result, many companies from a variety of industries are interested in the idea and have already engaged in projects using the metaverse that are especially targeted at the market for their individual industry (Du et al., 2022). One of the significant advantages of the metaverse for businesses is its open nature, as it is not owned or exclusive to any particular individual, group, nation, or business. This openness makes it easier for businesses to implement original ideas and innovations. Consequently, it is logical for companies across various industries to explore, develop, and provide a wide range of goods and services relevant to the metaverse in order to cater to their customers' needs. The metaverse offers a multitude of functionalities that can be applicable to clients from diverse industries. According to Liederman's (2021) research, the metaverse market is estimated to be valued at approximately US \$12.5 trillion. The presence of many metaverse tools exacerbates the problem of determining the best tool because many factors intervene to determine the best tool, such as privacy, security, and so on. Moreover, this study serves practitioners and academics by identifying the best metaverse tool based on the combination of the FWZIC method and the extended ARAS method using the Hexagonal Fuzzy Numbers.

2 Theoretical Background

The metaverse refers to a virtual, shared, and three-dimensional environment where users can engage in real-time online social and professional interactions, utilizing various technologies (Park and Kim, 2022; Eneizan et al., 2019). Shen et al. (2021) and Dalgarno and Lee (2010) suggest that the metaverse offers potential for enhancing the dissemination of spatial information, enabling the execution of experiential tasks that may not be feasible in the physical world, facilitating the creation of presence and identity, and improving the transfer of information to real-world contexts. The emergence of metaverse platforms like Fortnite, Roblox, and Minecraft exemplifies the convergence of virtual and tangible experiences across different domains such as fashion, music, education, and advertising. Through avatars and smart devices like smartphones and virtual reality headsets, users can fully immerse themselves in this virtual digital environment. Interactions within this virtual environment closely resemble real-life interactions, opening up a realm of possibilities (Cheong, 2022; Albahri et al., 2022). For instance, the industry will benefit from the metaverse by having a better means to plan future development. As a result, most companies (such as Amazon, Google, Apple, and others) are investing heavily in incorporating metaverse technologies into their current business. Many of these businesses are changing their business models to take advantage of opportunities in this virtual environment (e.g., marketing mix, service delivery). Digital evolution is considered a vital factor in the metaverse process (Vidal-Tomás, 2022).

The metaverse is the next step in the evolution of the digital age and has the potential to drastically alter how we interact with technology (Cheng et al., 2022). When compared to more traditional online-only platforms, it offers a wider variety of services. Services in the business, entertainment, and educational sectors, as well as any other system that can be connected to the internet, have all benefited greatly from the digitization trend of the previous few decades. Clients will be able to take immersive virtual tours of hotels to see the interiors, facilities, rooms, and amenities, before making their travel arrangements. There is a relationship between luxury brands and metaverse (Buhalis et al., 2022). To contribute to the creation of brand value, consumers engage with and experience brands (Lee et al., 2023). Emotional experience is a crucial component in producing such value (Arici et al., 2023). The conventional understanding of consumer behavior focuses on the logical, cognitive, and informational aspects. However, Holbrook and Hirschman (1982) argue that a solely information-based, rational approach to consumption is inadequate in explaining consumer behavior. They propose a different perspective that considers the intrinsic value of a product through subjective experiences encompassing sensory perceptions, aesthetic appeal, enjoyment, fantasy, and emotions. Customers can fully immerse themselves in the exhibitions, products, and culture of premium brands in the metaverse (Park and Kim, 2022). In other words, users of the metaverse system are more likely to stick around if they have positive interactions with good immersion (Jiang et al., 2023). Metaverse has played vital role during the COVID-19 pandemic (Pfefferbaum and North, 2020; Alnoor et al., 2022). Amidst the COVID-19 pandemic, individuals encountered unfavorable affective states such as depression and anxiety (Cheah and Shimul, 2023; Khair and Malhas, 2023) and resorted to compensation techniques to alleviate them (Golemis et al., 2022). Compensatory consumption, specifically revenge consumption, has been observed as a phenomenon in which individuals engage in increased consumption of goods and services as a means of compensation (Achille and Zipser, 2020). As confirmed by the findings of Pang et al. (2022), the COVID-19 pandemic has been related to a significant increase in prestige brand market sales. Jung et al. (2021) have observed that companies are facing limitations with web 2.0 in the wake of the COVID-19 pandemic, leading to a surge in demand for more streamlined platforms. In response to the deceleration in sales because of the COVID-19 pandemic, luxury and service enterprises have endeavored to establish closer relationships with their clientele and implement a novel business model that facilitates sustainable growth (Son et al., 2023). Luxury businesses give their customers access to the metaverse, where they enjoy utilizing fashion accessories and engaging with other avatars while feeling disconnected from reality (Cheah and Shimul, 2023). The metaverse supports the Holbrook and Hirschman (1982) experiential method to consuming, as well as user behaviors and real-time interactions. Consumers choose the items of a luxury brand based on desires and pleasures. Through brand identification, such as logos, colours, and sizes, luxury brands enable consumers to feel good (Khair and Malhas, 2023).

This entails engaging in fantasy within a metaverse setting that is distinct from the actual life. As a consequence, the term metaverse experience describes how customers connect with luxury companies through the goods and services they receive in virtual settings. Customers may search for luxury companies that give out good emotional cues

(Wang et al., 2022). Customers can engage in emotional and symbolic experiences while consuming a variety of information in a virtual arena given by luxury companies (Makkar and Yap, 2018; Muhsen et al., 2023a, b; Al-Hchaimi et al., 2023; Chew et al., 2023). The public just recently became aware of the metaverse, even though technology has evolved greatly, when Facebook modified its name to meta and disclosed that it was beginning to build its own versions of this technology (Kim and Bae, 2023). The metaverse is considered a vital factor in digital development (Lv et al., 2022). The metaverse, which broadens the selection of services available beyond conventional systems with online access, is the next stage of digital development and has the potential to transform digital adoption to unbelievable proportions (Mystakidis, 2022). In the last few decades, digitizing services has become a popular way to make them more productive in business, education, entertainment, and any other system that can be linked to the Internet (Abbas et al., 2023; Bozanic et al., 2023).

Digital systems and the ability to process and store data online at faraway data centers and cloud platforms made it possible for these systems and services to exist. There is a connection between online shopping and metaverse (Swilley, 2015). Consumers cannot touch or test the things they see online when they conventionally buy online (Zhang et al., 2023a, b, c). The metaverse is advantageous for industries including retail, fashion, gaming, sports, hospitality, and tourism (Hudson, 2022). Numerous businesses have already invested in the metaverse to allow buyers to visualise things in an immersive environment, including IKEA, Gucci, ZARA, and Alibaba among others (Joy et al., 2022). Programmers and designers are developing interactive devices to link retailers and customers in this fantastic world of shopping. Users will be able to make purchases while immersed in this digital space because of the presence of integrated branding and NFTs sales (Sayem, 2022). Metaverse is considered a crucial factor in the education process (Hwang & Chien, 2022). Students can develop and design in their own unique ways in the classroom as makers. Using diverse tools and the acquisition of new skills, they create new creative habits (Contreras et al., 2022). The respect and application of both new and ancient technologies is one of the characteristics of maker education that stands out the most. From this angle, numerous scholars have anticipated a rise in the usage of digital technology and its applications to create important and shareable content both offline and online (Inceoglu et al., 2022; Khaw et al., 2022).

For instance, Sun (2022), investigated how Online education and technology were combined to assess college students' perspectives. Researchers and academics have found that making use of online spaces effectively is important since it allows for more organized communication based on synchronous and asynchronous shared participation. According to other studies (Blum-Ross et al., 2019; Carretero et al., 2017), students in education can actively and experientially create their knowledge through digital competence. In a similar spirit, Cohen et al. (2017), underline the importance of digital technology in giving students the chance to recreate and forward other people's works with the maker community in addition to sharing their own. Metaverse is considered a critical factor in the healthcare process (Musamih et al., 2022). The healthcare industry may change the ecosystem as Metaverse-related technologies develop and are used (Moztarzadeh et al., 2023). A digital representation of reality exists in the metaverse. Massive volumes of data are produced to provide a more realistic and comprehensive experience (Bansal et al., 2022). The capacity of computers and data storage in the physical world has a major bearing on the scale and complexity of the metaverse.

The privacy and security of patient data is a significant concern for Metaverse healthcare systems. Telemedicine, remote patient care, diagnostics, and monitoring are all possible applications for the Metaverse (Ali et al., 2023). Space and temporal limits have been abolished in Metaverse healthcare systems using artificial intelligence, big data, virtual reality, other technologies, and augmented reality (Wiederhold & Riva, 2022). Metaverse healthcare systems gather medical data via wearable AI sensors. To ensure the security of healthcare data stored in the Metaverse, AI devices must locally encrypt medical data after collection, which means users will no longer have access to the plaintext. To achieve fine-grained control over the metaverse's healthcare data, attribute-based encryption is a popular cryptographic primitive (Zhang et al., 2023a). There is a relationship between artificial intelligence and metaverse (Hwang & Chien, 2022). Recent growth has influenced corporate technology advancements due to the advent of new technologies such as cloud computing, AI, and machine learning employed for data extraction goals (Huynh et al., 2023). AI is the intelligence demonstrated by machines as opposed to human intellect (Hussain et al., 2019). According to Verma (2021), AI can be defined as computers that can activate cognitive and emotive functions that are comparable to human abilities. In complicated situations and from several perspectives, rational decision-making is made possible because of the convergence of computer technologies (Akter et al., 2020). In addition, AI refers to a collection of techniques that enable robots to think, reason, and act in ways that are comparable to those of humans (De Bruyn et al., 2020).

According to Kushwaha et al. (2021), the ability to identify patterns in data and automate tasks has progressed to the extent that computers are now capable of making predictions. Through the automated replication of virtual actions in the physical world, AI facilitates the generation of digital replicas in the metaverse (Barrera & Shah, 2023). This section describes the relationship between metaverse and environmental (Sa et al., 2023). The potential impact of the metaverse on environmental sustainability is a topic of contentious discussion, as scholars have yet to reach a consensus on whether it will have a detrimental or beneficial effect on the atmosphere (Dwivedi, et al., 2022). According to one perspective, the metaverse enables virtual work, collaboration, shopping, and study, that could eliminate physical activities like commuting for work and reduce energy consumption (Beck et al., 2023). According to Ryu et al. (2022), in a multinational professional services company, it will be possible to achieve savings in energy consumption and costs of up to 50% by making a digital twin of a building in the metaverse and including artificial intelligence and machine learning (Zallio & Clarkson, 2022). Sustainability is considered a vital factor in the metaverse (Zhang et al., 2023a, b, c). There exist multiple interrelations between the metaverse and sustainability. Given the anticipated significant influence of the metaverse on various domains, including the material economy, social fabric, and environmental sustainability at a global scale, the issue of sustainability assumes paramount importance.

The depletion of crucial natural resources is a likely outcome of escalating consumption and population growth unless they are managed prudently and recycled effectively.

While the metaverse would boost economic growth and financial success, these elements would have to be balanced against the metaverse's environmental, social, and ethical aspects. (Jauhiainen et al., 2023). Many technological development businesses place a high priority on sustainability objectives that reduce their present carbon emissions through more energy-efficient operations. There is a relationship between metaverse and innovation (Kwok et al., 2023). One of the key aspects in customer-focused design strategy is the product enclosure design, as the visual appeal of a product often plays a significant role in defining its concept innovation. Despite extensive research on customer-focused innovation and psychological product design, creating a new and imaginative product that resonates with the target customers remains challenging (Yung et al., 2021). Understanding customer needs and incorporating desired product functionalities can be complex tasks. Furthermore, achieving a balance between enhancing the technical aspects and the aesthetic appeal of a product can be a daunting challenge.

3 Methodology and Results

Our methodology is divided into 3 sections: the first is the decision matrix, the second is FWZIC, and the third is ARAS.

Firstly, the decision matrix consists of a set of alternatives (10 tools), three main criteria, and 18 sub-criteria. On the one hand, the alternatives are the control engineering tools used for supporting ICPMSMS are researched and identified. According to the findings of the research that was presented, ten different tools have been identified here. Siemens-TIA-Portal-Openness and-SiVArc (Tool_1), Mitsubshi_Adroit_Process_Suite (Tool_2), Schneider_Electric (Tool_3), Beckhoff_TwinCAT_Automation_Interface (tool_4), General_Electric (Tool_5), Rockwell_Automation (Tool_6), Bosch (Tool_7), MATLAB_Simulink_PLC-Coder (Tool_8), DELMIA (Tool_9), and WinMOD (Tool_10).

On the other hand, in order to assess the chosen IOC control engineering tools, we analyse and catalogue the ICPS components. There are four criteria by which these key parts of an ICPS include the PLC and HMI, cybersecurity, and digital twin. Each ICPMSMS-supporting tool must also fulfil all subcriteria within the three ICPS key elements. PLC hardware generation (C_1), PLC tags generation (C_2), PLC software library generation (C_3), PLC code generation (C_4), HMI screen generation (C_5), interoperability (C_6), PLCopen XML generation (C_7), and IEC61499 programming language generation (C_8) are the eight criteria that make up the PLC and HMI criteria. All PLC and HMI criteria are broken down into eight subcriteria that must be met by any tool used to enable ICPMSMS. In addition, OPC-UASecurity-Controls (C_1), Online-Code-Protection (C_2), Know-How Protection (C_3), Access Control(C_4), Hardware-Security-Protection (C_5), and SoftwareChanges Trailing (C_6) are the six subcriteria that make up the cybersecurity criterion. The digital twin criterion consists of six subcriteria, namely, Component-Based-Modeling (C_1), VES Modeling including Dynamic Environment Simulation (C_3), Real-Time-Simulation (C_4), and Dynamic Environment Simulation (C_2). Each assisting tool for ICPMSMS must meet these four subcriteria in order to fulfil the digital twin requirement. For more information on the primary and supplementary ICPS parts.

Secondly, we will go into detail on the five basic stages that make up the Hexagonal-FWZIC approach.

The Initial Stage: The set of evaluations definition. At this stage, we define a set of evaluation criteria that will be further elaborated upon when we proceed to the expert evaluation phase.

Second Stage: Expert selection. The importance of the criteria is evaluated using a Structured Expert Judgment (SEJ) procedure in which experts from the relevant field take part. This stage consists of five sub-steps. Experts in the field are sought out first, and then a panel of at least three is assembled for each area of study. However, studies show that an expert group of 3–5 persons is most effective (Albahri et al. 2022; Muhsen et al. 2023a, b). The third sub-step involves creating an assessment model critical for gathering data. The experts chosen in Step 1 evaluate the reliability and validity of the questionnaire. In the fourth phase, the importance of each criterion is determined using a five-point Likert scale, which helps reduce bias and guarantees a high level of dependability. Finally, the linguistic scale is transformed into a similar numerical scale. Table 1 displays the numerical values assigned to each expert.

Table 1. Five-point Likert Scale.

Linguistic scoring scale	Numerical scoring scale
Not-important	1
Slight-important	2
Moderately-important	3
Important	4
Very-important	5

Third Stage: The process of building an Expert Decision Matrix (EDM) involves creating a matrix that incorporates the alternatives and criteria outlined in Table 2. This matrix illustrates the relationship between each criterion (Cj) and the selected experts (Ei), where each expert assesses the level of importance for each criterion.

Table 2. EDM.

Criteria	$C1$	$C2$	...	Cn
$E1$	$(E1/C1)$	$(E1/C2)$	...	$(E1/Cn)$
$E2$	$(E2/C1$	$(E2/C2)$	...	$(E2/Cn)$
...	...	...	...	...
Em	$(En/C1)$	(En/C2)	...	(Em/Cn)

Step four involves the utilization of a hexagonal membership function. This stage applies a fuzzy function to the membership and defuzzification process using the data from the EDM. By employing this approach, the accuracy and utilization of information are enhanced, which is particularly important as MCDM faces challenges related to rigid preferences for each criterion. To address this issue, fuzzy numbers are used instead of crisp numbers to determine the criterion values (refer to Table 3). The hexagonal set, defined by definitions 1 and 2, is employed as an objective form.

Table 3. Linguistic Terms.

Linguistic terms	HFN
Not-important	(1, 2, 4, 6, 7, 9)
Slight-important	(2, 4, 6, 7, 9, 11)
Moderately-important	(4, 6, 7, 9, 11, 13)
Important	(6, 7, 9, 11, 13, 15)
Very-important	(7, 9, 11, 13, 15, 16)

A Hexagonal Fuzzy membership function is given below (Parveen and Kamble 2020; Sudha and Revathy 2016):

$$
\begin{cases}
\frac{1}{2}\left(\frac{x-f_1}{f_2-f_1}\right), & f_1 \leq x \leq f_2 \\
\frac{1}{2}+\frac{1}{2}\left(\frac{x-f_2}{f_3-f_2}\right), & f_2 \leq x \leq f_3 \\
1, & f_3 \leq x \leq f_4 \\
1-\frac{1}{2}\left(\frac{x-f_4}{f_5-f_4}\right), & f_4 \leq x \leq f_5 \\
\frac{1}{2}\left(\frac{f_6-x}{f_6-f_5}\right), & f_5 \leq x \leq f_6 \\
0, & \text{otherwise}
\end{cases}
\tag{1}
$$

Definition (1)

A fuzzy set $\tilde{F}\_H$ an be defined as a collection of real numbers, where its membership function possesses the following characteristics (Sudha and Revathy 2016).

Function P1(u) is an ongoing function which is restricted on the left side and does not lower over the interval [0, 0.5].

The function Q1(v) is an ongoing function which is restricted on the left side and does not lower over the interval [0.5, w].

The function Q2(v) is an ongoing function that is restricted on the left side and does not raise over the interval [w, 0.5].

The function P2(v) is an ongoing function that is restricted on the left side and does not raise over the interval [0.5, 0].

NOTE: If the value of w is equal to 1, the fuzzy hexagonal number becomes a standard fuzzy number.

Definition (2)

Operation on Hexagonal Fuzzy Numbers:

The main for two Hexagonal Fuzzy Numbers (HFN) U = (u-1, u-2, u-3, u-4, u-5, u-6) and V = (v-1, v-2, v-3, v-4, v-5, v-6) can be defined as follows: (Ghosh et al. 2021), (Parveen and Kamble 2020), (Nayagam, Murugan, and Suriyapriya 2020):

Addition:

$$(U + V) = (u_1 + v_1, \ldots, u_6 + v_6) \tag{2}$$

Subtraction:

$$(U - V) = (u_1 - v_6, \ldots, u_6 - v_1) \tag{3}$$

Multiplication:

$$(U \times V) = (u_1 v_1, \ldots, u_6 v_6) \tag{4}$$

Division:

$$\left(\frac{U}{V}\right) = \left(\frac{u_1}{v_6}, \ldots, \frac{u_6}{v_1}\right) \tag{5}$$

Fifth Step. Final values of the weight calculation. In this stage, the assessment criteria's weight coefficients are determined. Fuzzification data from the previous stage is used to figure out the following values for the weight coefficients of the assessment criteria: $(w1, w2, \ldots, wn)T$.

(i) To determine the ratio of fuzzification data using Eqs. (6).

$$\frac{\mathrm{Imp}(\widetilde{E1}/C1)}{\sum_{j=1}^{n} \mathrm{Imp}(\widetilde{E1}/C_{1j})} \tag{6}$$

where $\mathrm{Imp}\left(\frac{\widetilde{E11}}{C1}\right)$ represent the fuzzy number of $\mathrm{Imp}\left(\frac{E1}{C1}\right)$

(ii) To obtain the final fuzzy values of the weight coefficients for the evaluation criteria $(\widetilde{w}, \widetilde{w2}, \ldots, \widetilde{wn})^T$ the mean values are calculated. The Fuzzy EDM $\left(\widetilde{\mathrm{EDM}}\right)$ is used for calculating the final weight number for every criterion through Eq. (7).

$$\widetilde{W_J} = \left(\sum_{i=1}^{m} \frac{\mathrm{Imp}\overline{(E_{tj}/C_{tj})}}{\sum_{j=1}^{n} \mathrm{Imp}(E(E_{tj}/C_{tj}))}\right)/m, \quad \text{for } i = 1, 2, 3, \ldots m \text{ and } j = 1, 2, \ldots n. \tag{7}$$

Fuzzy weights $\widetilde{w_J}$ are assigned to each criterion by dividing their ratio values by the total number of experts, $Imp(E1/\widetilde{C}1)$ represent the fuzzy number of Imp (E1/C1), and $\sum_{j=1}^{n} \mathrm{Imp}(E\left(\frac{E_{tj}}{C_{tj}}\right))$ is the summation of all fuzzy number values of the importance assigned by the expert per criterion. Then the defuzzification is figured via Eq. (8) (Ghosh et al. 2021):

$$= \left(\frac{3h_1 + 3h_2 + 10h_3 + 10h_4 + 5h_5 + 3h_6}{34}\right) \tag{8}$$

Thirdly, Zavadskas and Turskis proposed a novel MCDM approach called (ARAS), which is both efficient and simple to implement. ARAS contents 5 steps will be explained below (Toygar, Yildirim, and İnegöl 2022):

Step 1: Identify the optimum assessment of performance for every criterion.

After building a decision matrix, the next phase in the ARAS technique is to identify the best possible rating for every criterion. If experts have no preferences, the most effective performance ratings are determined as:

$$x_{0j} = \begin{cases} \max_i x_{ij}; \ j \in \Omega_{\max} \\ \min_i x_{ij}; \ j \in \Omega_{\min} \end{cases} \tag{9}$$

where x_{0j} represents the best possible rating with respect to criterion j-th criterion, $\Omega_{\max}$ stands for a set of benefit-type criteria (maximisation); and $\Omega_{\min}$ min stands for a set of cost-type criteria (minimisation).

Step 2: Determine the normalised decision matrix $R = \left[r_{ij} \right]$. Formulas for calculating normalised performance evaluations are as follows:

$$r_{ij} = \begin{cases} \frac{x_{ij}}{\sum_{i=0}^{m} x_{ij}}; \ j \in \Omega_{\max} \\ \frac{1/x_{ij}}{\sum_{i=0}^{m} 1/x_{ij}}; \ j \in \Omega_{\min} \end{cases}, \tag{10}$$

where r_{ij} is the normalised of i-th alternative about the j-th criterion.

Step 3: The weighting of the decision matrix $V = \left[v_{ij} \right]$. The following formula is used to get weighted adjusted performance rankings:

$$v_{ij} = w_j \cdot r_{ij}; \quad i = 1, 2, \ldots, m \tag{11}$$

where v_{ij} is a weighted decision matrix.

Step 4: Find the total efficacy index for every alternative. The overall effectiveness index S_i, can be computed for each alternative utilising a formula as follows:

$$S_i = \sum_{j=1}^{n} v_{ij}; \ i = 0, 1, \ldots, m \tag{12}$$

Step 5: Calculate the final rank. In the context of assessing metaverse applications, the following formula can be used to determine the degree of effectiveness:

$$Q_i = \frac{S_i}{S_0}; \quad i = 1, 2, \ldots, m \tag{13}$$

where Q_i is the mark of the effectiveness of i-th alternative, and S_0 is the summation of optimal alternative. Keep in mind that the largest value of Q_i is the top.

By using Table 2, the opinion of five experts specialized in the field of the metaverse is collected to produce the EDM as shown in Table 4 for the main criteria.

Table 4. Crisp number.

Exp1	4	4	3
Exp2	4	5	4
Exp3	3	3	3
Exp4	4	5	5
Exp5	4	4	4

After collecting the fuzzy EDM, we applied the aforementioned Eqs. (1–8) in the first step of our methodology to get the weight, the results for each of the main and sub-criteria and the final weights as shown in Table 5 below.

Table 5. Weighting Results in Icps Components.

Main criteria	Weight	Sub-criteria	Sub-criteria	Final
c1	0.328	c1_1	0.130	0.042
		c1_2	0.116	0.038
		c1_3	0.138	0.045
		c1_4	0.130	0.042
		c1_5	0.120	0.039
		c1_6	0.121	0.039
		c1_7	0.115	0.037
		c1_8	0.126	0.041
c2	0.348	c2_1	0.147	0.051
		c2_2	0.167	0.058
		c2_3	0.175	0.060
		c2_4	0.170	0.059
		c2_5	0.165	0.057
		c2_6	0.173	0.060
c3	0.325	c3_1	0.236	0.076
		c3_2	0.243	0.079
		c3_3	0.258	0.083
		c3_4	0.260	0.084

Table 5 shows the local and final weights of the primary and sub-ICPS components. In terms of the main criteria, c2 received the highest weight of 0.347, while c3 obtained the lowest weight of 0.324. In addition, according to the final weight, c3-4 had the highest weight of 0.084 and c1-7 was the lowest one with a weight of 0.037.

After obtaining the weight of all criteria, the ARAS is applied to get the final rank of alternatives.

In the context of the present study, where the assessment of metaverse tools is based on a privacy model using the fuzzy MCDM method, the highest possible score signifies the best choice for each criterion. The following step is to calculate the normalized weight by multiplication of the normalized matrices with the criteria weights. Table 6 is shown in the score matrix. The normalized weight is calculated by multiplying the normalized matrices by the supplied criteria weights in this essential phase. This procedure ensures that the importance of each criterion is correctly accounted for in the evaluation of metaverse tools.

Table 6. Collect all the Values for each Alternative.

Alternatives	Score
S0	0.240
Tool_1	0.207
Tool_2	0.090
Tool_3	0.095
Tool_4	0.095
Tool_5	0.032
Tool_6	0.053
Tool_7	0
Tool_8	0.043
Tool_9	0.083
Tool _10	0.055

The table offers an assessment of various alternatives based on a set of criteria, and the "Score" column displays the appropriate scores provided to each tool. The scores contain normalized values between 0 and 1, with 1 representing the best score and 0 representing the worst score for each criterion. The fifth step is calculating the final rank, where the value of each alternative is divided by S0 to produce a final rank. Table 7 shows the final rank.

Table 7 illustrates the evaluation of several tools based on a specified set of criteria, with each tool receiving a "Score" and a "Final Rank" based on how well it performed against those criteria. The "Score" column provides values indicating how well each tool fared in the evaluation, and the "Final Rank" column displays each tool's ranking based on its score. Tool_1 received the highest score of 0.861, indicating that it was the best-performing tool in the evaluation, ranking first. Furthermore, Tool_3 received a score of 0.398 and finished second in the ranking. Tool_7, on the other hand, received a 0 and was ranked last, in tenth place.

Table 7. Final Rank.

Tools	Score	Final Rank
Tool_1	0.861	1
Tool_2	0.374	4
Tool_3	0.398	2
Tool_4	0.396	3
Tool_5	0.136	9
Tool_6	0.222	7
Tool_7	0	10
Tool_8	0.180	8
Tool_9	0.348	5
Tool _10	0.232	6

4   Evaluation

Concerning the ability to be generalized findings require immediate intervention, which can be accomplished through evaluation. This study used objective validation procedures to ensure that a rigorous classification technique was used to rank the metaverse tools. Notably, the number of groups or alternatives will have no impact on the process of assessment. For the systematic ranking, the following stages are taken: (1) The tools are ordered based on the end results. (2) The groups are divided equally after sorting. (3) The evaluation results are then calculated using the mean (x) for each group, as shown below in Table 8. The 1st group's mean should be greater than or equal to the mean of the 2nd group.

Table 8. Evaluation Result.

No:	Groups	Mean Result
1	1ST Group	0.4759
2	2ND Group	0.1542

In this context, the table compares the results of two groups, referred to as the "1st Group" and the "2nd Group." Each group's average or mean value is represented in the "Mean Result" column. The average result for the "First Group" is 0.4759. The "2nd Group" on the other hand, received a mean score of 0.1542. When the mean values are compared, we can see that the first group outperformed the second group, with a substantially higher mean value of 0.4759 compared to 0.1542. This research demonstrates the validity of ranking and benchmarking metaverse tools.

5 Managerial Implications

The use of the fuzzy MCDM approach to evaluate metaverse tools based on a privacy model has essential managerial implications. Organizations can assure increased data privacy compliance, eliminate risks and manage reputation, optimize resource allocation, and easily integrate technologies into existing systems by implementing this methodology. Furthermore, the approach prioritizes user experience and adoption, providing a competitive edge via privacy-focused solutions. Long-term strategic planning is made easier, allowing for continual improvement and adaptability to changing privacy and security concerns in the metaverse landscape, resulting in a resilient and privacy-conscious firm. Finally, our methodology allows managers to measure and analyze numerous characteristics, such as data privacy features, usability, scalability, and cost, to determine which metaverse technologies provide the best return on investment.

6 Conclusion

The metaverse has revolutionized the business world and changed the technology landscape around the world. With metaverse technology, there are many technologies that emerge with multiple tools; metaverse technology increases the difficulty of determining the best alternative usable by companies and users. For example, in light of the issue of privacy and security protection, determining the best alternative is a multi-criteria decision problem. In addition, this study identified ten metaverse tools. To this end, this study aims to adopt multi-criteria decision-making methods extended by the fuzzy method to determine the best tool for metaverse technology on the basis of privacy and security. The FWZIC expanded by the Hexagonal Fuzzy Numbers method was utilized to determine the importance of criteria for classifying metaverse technology tools. However, the FWZIC method has the inability to rank the tools from best to worst. Hence, this study used the ARAS method to rank the tools from best to worst based on the importance of criteria. According to the outcome of the Hexagonal-FWZIC analysis, cybersecurity is the most influential factor, while the digital twin-component has the least. The ARAS findings indicate that, of the ten tools studied, tools_1, tool_3, and tool_4 are the best three tools. This study was distinguished from the previous literature by defining the best metaverse tool with high certainty, where the fuzzy type used is considered more certain compared to other types. Therefore, the current study addressed the problem of ambiguity in decision-making. The current study provides theoretical and practical contributions for academics and practitioners. Theoretically, expanding the theories and methods of decision-making with fuzzy types opens many doors and horizons for literature and academics to develop decision-making methods and address the problems of ambiguity and uncertainty. Practically identifying the best-performing metaverse technology gives practitioners ample opportunities to develop more successful virtual projects based on the most successful tools in accordance with expanded decision-making methods. This investigation contributes to Industry 5.0 in order to further develop applications and address trustworthiness problems in many virtual domains. Thus, this study is a contribution to the field of increasing the trustworthiness of the used virtual purifications. Nevertheless, our study is in line with previous literature, with limitations that require immediate intervention by future literature and existing research. The number of experts can be increased to evaluate the decision matrix and evaluate each expert before a weight is recorded for the criteria. Also, the use of other fuzzy types is important to address the problems of ambiguity in decision-making. Finally, we suggest future studies to study customers' perceptions of the best metaverse technology tools on the basis of trustworthiness using trust theory, social theories, and technology acceptance theories.

References

Abbas, S., et al.: Antecedents of trustworthiness of social commerce platforms: a case of rural communities using multi group SEM & MCDM methods. Electron. Commer. Res. Appl. **62**, 101322 (2023)

Achille, A., Zipser, D.: A perspective for the luxury-goods industry during and after coronavirus. McKinsey Insights, pp. 1–8, 01 April 2020

Akter, S., Gunasekaran, A., Wamba, S.F., Babu, M.M., Hani, U.: Reshaping competitive advantages with analytics capabilities in service systems. Technol. Forecast. Soc. Chang. **159**, 120180 (2020)

Albahri, O.S., et al.: Novel dynamic fuzzy decision-making framework for COVID-19 vaccine dose recipients. J. Adv. Res. **37**, 147–168 (2022)

AL-Fatlawey, M.H., Brias, A.K., Atiyah, A.G.: The role of Strategic Behavior in achievement the Organizational Excellence "Analytical research of the manager's views of Ur State Company at Thi-Qar Governorate". J. Admin. Econ. **10**(37) (2021)

Al-Hchaimi, A.A.J., Sulaiman, N.B., Mustafa, M.A.B., Mohtar, M.N.B., Hassan, S.L.B.M., Muhsen, Y.R.: A comprehensive evaluation approach for efficient countermeasure techniques against timing side-channel attack on MPSoC-based IoT using multi-criteria decision-making methods. Egypt. Inf. J. **24**(2), 351–364 (2023)

Ali, S., et al.: Metaverse in healthcare integrated with explainable AI and blockchain: enabling immersiveness, ensuring trust, and providing patient data security. Sensors **23**(2), 565 (2023)

Alnoor, A., et al.: How positive and negative electronic word of mouth (eWOM) affects customers' intention to use social commerce? A dual-stage multi group-SEM and ANN analysis. Int. J. Hum. Comput. Interact., 1–30 (2022)

Arici, H.E., Saydam, M.B., Koseoglu, M.A.: How do customers react to technology in the hospitality and tourism industry? J. Hosp. Tour. Res., 10963480231168609 (2023)

Atiyah, A.G.: Effect of temporal and spatial myopia on managerial performance. J. La Bisecoman **3**(4), 140–150 (2022)

Bansal, G., Rajgopal, K., Chamola, V., Xiong, Z., Niyato, D.: Healthcare in metaverse: a survey on current metaverse applications in healthcare. IEEE Access **10**, 119914–119946 (2022)

Barrera, K.G., Shah, D.: Marketing in the Metaverse: conceptual understanding, framework, and research agenda. J. Bus. Res. **155**, 113420 (2023)

Beck, D., Morgado, L., O'Shea, P.: Educational practices and strategies with immersive learning environments: mapping of reviews for using the Metaverse. IEEE Trans. Learn. Technol. (2023). https://doi.org/10.1109/TLT.2023.3243946

Blum-Ross, A., Kumpulainen, K., Marsh, J.: Enhancing Digital Literacy and Creativity. Routledge, London, UK (2019)

Bozanic, D., Tešić, D., Puška, A., Štilić, A., Muhsen, Y.R.: Ranking challenges, risks and threats using Fuzzy Inference System. Decis. Making Appl. Manage. Eng. **6**(2), 933–947 (2023)

Buhalis, D., Lin, M.S., Leung, D.: Metaverse as a driver for customer experience and value co-creation: implications for hospitality and tourism management and marketing. Int. J. Contemp. Hosp. Manag. **35**(2), 701–716 (2022)

Carretero, S., Vuorikari, R., Punie, Y.: DigComp 2.1: the digital competence framework for citizens (2017)

Cheah, I., Shimul, A.S.: Marketing in the Metaverse: moving forward–what's next? J. Glob. Scholars Market. Sci. **33**(1), 1–10 (2023)

Cheng, X., et al.: Exploring the metaverse in the digital economy: an overview and research framework. J. Electron. Bus. Digit. Econ. **1**(1/2), 206–224 (2022)

Cheong, B.C.: Avatars in the Metaverse: potential legal issues and remedies. Int. Cybersecur. Law Rev. **3**(2), 467–494 (2022)

Chew, X., Khaw, K.W., Alnoor, A., Ferasso, M., Al Halbusi, H., Muhsen, Y.R.: Circular economy of medical waste: novel intelligent medical waste management framework based on extension linear Diophantine fuzzy FDOSM and neural network approach. Environ. Sci. Pollut. Res. **30**, 60473–60499 (2023)

Cohen, J.: Maker principles and technologies in teacher education: a national survey. J. Technol. Teach. Educ. **25**(1), 5–30 (2017)

Contreras, G.S., González, A.H., Fernández, M.I.S., Martínez, C.B., Cepa, J., Escobar, Z.: The importance of the application of the Metaverse in education. Mod. Appl. Sci. **16**(3), 1–34 (2022)

Dalgarno, B., Lee, M.J.W.: What are the learning affordance of 3-D virtual environment? Br. J. Edu. Technol. **41**(1), 10–32 (2010)

De Bruyn, A., Viswanathan, V., Beh, Y.S., Brock, J.K.U., Von Wangenheim, F.: Artificial intelligence and marketing: pitfalls and opportunities. J. Interact. Mark. **51**(1), 91–105 (2020)

Dionisio, J.D.N., Burns, W.G., III., Gilbert, R.:3D virtual worlds and the Metaverse: current status and future possibilities. ACM Comput. Surv. **45**(3), 1–38 (2013)

Du, H., Niyato, D., Miao, C., Kang, J., Kim, D.I.: Optimal targeted advertising strategy for secure wireless edge metaverse. In:2022 IEEE Global Communications Conference, pp. 4346–4351, GLOBECOM 2022. IEEE, December 2022

Dwivedi, Y.K., et al.: Metaverse beyond the hype: multidisciplinary perspectives on emerging challenges, opportunities, and agenda for research, practice and policy. Int. J. Inf. Manage. **66**, 102542 (2022)

Dwivedi, Y.K., et al.: Metaverse marketing: how the Metaverse will shape the future of consumer research and practice. Psychol. Mark. **40**(4), 750–776 (2023)

Eneizan, B., Mohammed, A.G., Alnoor, A., Alabboodi, A.S., Enaizan, O.: Customer acceptance of mobile marketing in Jordan: an extended UTAUT2 model with trust and risk factors. Int. J. Eng. Bus. Manage. **11**, 1847979019889484 (2019)

Gatea, A.A., Marina, V.: Higher education funding in Iraq in terms of the experience of particular developed countries. Int. J. Adv. Stud. **6**(1), 8–17 (2016)

Ghosh, A., et al.: Application of hexagonal fuzzy MCDM methodology for site selection of electric vehicle charging station. Mathematics **9**(4), 393 (2021)

Golemis, A., et al.: Young adults' coping strategies against loneliness during the COVID-19-related quarantine in Greece. Health Promot. Int. **37**(1), 1–13 (2022). daab053

Hamid, R.A., et al.: How smart is e-tourism? A systematic review of smart tourism recommendation system applying data management. Comput. Sci. Rev. **39**, 100337 (2021)

Holbrook, M.B., Hirschman, E.C.: The experiential aspects of consumption: consumer fantasies, feelings, and fun. J. Consum. Res. **9**(2), 132–140 (1982)

Hollensen, S., Kotler, P., Opresnik, M.O.: Metaverse – the new marketing universe. J. Bus. Strateg. **44**(3), 119–125 (2023)

Hudson, J.: Virtual Immersive shopping experiences in Metaverse environments: predictive customer analytics, data visualization algorithms, and smart retailing technologies. Linguist. Philos. Inv. **21**, 236–251 (2022)

Hussain, B., Du, Q., Imran, A., Imran, M.A.: Artificial intelligence-powered mobile edge computing-based anomaly detection in cellular networks. IEEE Trans. Industr. Inf. **16**(8), 4986–4996 (2019)

Huynh-The, T., Pham, Q.V., Pham, X.Q., Nguyen, T.T., Han, Z., Kim, D.S.: Artificial intelligence for the Metaverse: a survey. Eng. Appl. Artif. Intell. **117**, 105581 (2023)

Hwang, G.J., Chien, S.Y.: Definition, roles, and potential research issues of the Metaverse in education: an artificial intelligence perspective. Comput. Educ. Artif. Intell. **3**, 100082 (2022)

Inceoglu, M.M., Ciloglugil, B.: Use of Metaverse in education. In: Gervasi, O., Murgante, B., Misra, S., Ana, M.A., Rocha, C., Garau, C. (eds.) Computational Science and Its Applications – ICCSA 2022 Workshops: Malaga, Spain, 4–7 July 2022, Proceedings, Part I, pp. 171–184. Springer, Cham (2022). https://doi.org/10.1007/978-3-031-10536-4_12

Jauhiainen, J.S., Krohn, C., Junnila, J.: Metaverse and sustainability: systematic review of scientific publications until 2022 and beyond. Sustainability **15**(1), 346 (2023)

Jiang, Q., Kim, M., Ko, E., Kim, K.H.: The metaverse experience in luxury brands. Asia Pac. J. Mark. Logist. (2023, ahead-of-print)

Joy, A., Zhu, Y., Peña, C., Brouard, M.: Digital future of luxury brands: Metaverse, digital fashion, and non-fungible tokens. Strateg. Chang. **31**(3), 337–343 (2022)

Jung, J., Yu, J., Seo, Y., Ko, E.: Consumer experiences of virtual reality: insights from VR luxury brand fashion shows. J. Bus. Res. **130**, 517–524 (2021)

Khair, N., Malhas, S.: Fashion-related remedies: exploring fashion consumption stories during Covid-19. 'Nostalgia overpowering, Old is the new me.' J. Glob. Fash. Market. **14**(1), 77–92 (2023)

Khaw, K.W., et al.: Modelling and evaluating trust in mobile commerce: a hybrid three stage Fuzzy Delphi, structural equation modeling, and neural network approach. Int. J. Hum. Comput. Interact. **38**(16), 1529–1545 (2022)

Kim, J., Bae, J.: Influences of persona self on luxury brand attachment in the Metaverse context. Asia Pac. J. Market. Logist. (2023, ahead-of-print)

Kushwaha, A.K., Kumar, P., Kar, A.K.: What impacts customer experience for B2B enterprises on using AI-enabled chatbots? Insights from Big data analytics. Ind. Mark. Manage. **98**, 207–221 (2021)

Kwok, C.P., Tang, Y.M.: A fuzzy MCDM approach to support customer-centric innovation in virtual reality (VR) metaverse headset design. Adv. Eng. Inform. **56**, 101910 (2023)

Lee, N., Jo, M.: Exploring problem-based learning curricula in the Metaverse: the hospitality students' perspective. J. Hosp. Leis. Sport Tour. Educ. **32**, 100427 (2023)

Liederman, E.: As marketers enter the Metaverse, they must learn to 'think in 3D', Ad Week (2021). https://www.adweek.com/commerce/emerging-tech-tips-for-marketers/. Accessed 4 Dec 2021

Lv, Z., Xie, S., Li, Y., Hossain, M.S., El Saddik, A.: Building the Metaverse by digital twins at all scales, state, and relations. Virt. Reality Intell. Hardw. **4**(6), 459–470 (2022)

Makkar, M., Yap, S.F.: The anatomy of the inconspicuous luxury fashion experience. J. Fashion Market. Manage. Int. J. **22**(1), 129–156 (2018)

Moztarzadeh, O., et al.: Metaverse and healthcare: machine learning-enabled digital twins of cancer. Bioengineering **10**(4), 455 (2023)

Muhsen, Y.R., Husin, N.A., Zolkepli, M.B., Manshor, N., Al-Hchaimi, A.A.J.: Evaluation of the Routing Algorithms for NoC-based MPSoC: a fuzzy multi-criteria decision-making approach. IEEE Access **11**, 102806–102827 (2023)

Muhsen, Y.R., Husin, N.A., Zolkepli, M.B., Manshor. N.: A systematic literature review of fuzzy-weighted zero-inconsistency and fuzzy-decision-by-opinion-score-methods: assessment of the past to inform the future. J. Intell. Fuzzy Syst. **45**(3), 4617–4638 (2023)

Musamih, A., et al.: Metaverse in healthcare: applications, challenges, and future directions. IEEE Consum. Electron. Mag. **12**(4), 33–46 (2022)

Mystakidis, S.: Metaverse. Encyclopedia **2**(1), 486–497 (2022)

Nayagam, V.L.G., Murugan, J., Suriyapriya, K.: Hexagonal fuzzy number inadvertences and its applications to MCDM and HFFLS based on complete ranking by score functions. Comput. Appl. Math. **39**(4), 1–47 (2020)

Pang, W., Ko, J., Kim, S.J., Ko, E.: Impact of COVID-19 pandemic upon fashion consumer behavior: focus on mass and luxury products. Asia Pac. J. Mark. Logist. **34**(10), 2149–2164 (2022)

Park, S., Kim, S.: Identifying world types to deliver gameful experiences for sustainable learning in the metaverse. Sustainability **14**(3), 1361 (2022)

Parveen, N., Kamble, P.N.: Decision-making problem using fuzzy TOPSIS method with hexagonal fuzzy number. In: Iyer, B., Deshpande, P.S., Sharma, S.C., Shiurkar, U. (eds.) Computing in Engineering and Technology. AISC, vol. 1025, pp. 421–430. Springer, Singapore (2020). https://doi.org/10.1007/978-981-32-9515-5_40

Pfefferbaum, B., North, C.S.: Mental health and the Covid-19 pandemic. N. Engl. J. Med. **383**(6), 510–512 (2020)

Ryu, J., Son, S., Lee, J., Park, Y., Park, Y.: Design of secure mutual authentication scheme for metaverse environments using blockchain. IEEE Access **10**, 98944–98958 (2022)

Sá, M.J., Serpa, S.: Metaverse as a learning environment: some considerations. Sustainability **15**(3), 2186 (2023)

Sayem, A.S.M.: Digital fashion innovations for the real world and metaverse. Int. J. Fashion Des. Technol. Educ. **15**(2), 139–141 (2022)

Shen, B., Tan, W., Guo, J., Zhao, L., Qin, P.: How to promote user purchase in Metaverse? A systematic literature review on consumer behavior research and virtual commerce application design. Appl. Sci. **11**, 11087 (2021)

Smart, J., Cascio, J., Paffendorf, J.: Metaverse roadmap overview (2007). https://www.metaverse roadmap.org/overview/index.html. Accessed 10 Jul 2021

Son, S.C., Bae, J., Kim, K.H.: An exploratory study on the perceived agility by consumers in luxury brand omni-channel. J. Glob. Scholars Market. Sci. **33**(1), 154–166 (2023)

Sudha, A.S., Revathy, M.: A new ranking on hexagonal fuzzy numbers. Int. J. Fuzzy Logic Syst. **6**(4), 1–8 (2016)

Sun, M., et al.: The Metaverse in current digital medicine. Clin. eHealth **5**, 52–57 (2022)

Swilley, E.: Moving virtual retail into reality: examining Metaverse and augmented reality in the online shopping experience. In: Campbell, C., Ma, J. (eds.) Looking Forward, Looking Back: Drawing on the Past to Shape the Future of Marketing, pp. 675–677. Springer International Publishing, Cham (2016). https://doi.org/10.1007/978-3-319-24184-5_163

Toygar, A., Yildirim, U., İnegöl, G.M.: Investigation of empty container shortage based on SWARA-ARAS methods in the COVID-19 era. Eur. Transp. Res. Rev. **14**(1), 8 (2022)

Verma, S., Sharma, R., Deb, S., Maitra, D.: Artificial intelligence in marketing: systematic review and future research direction. Int. J. Inf. Manage. Data Insights **1**(1), 100002 (2021)

Vidal-Tomás, D.: The new crypto niche: NFTs, play-to-earn, and metaverse tokens. Financ. Res. Lett. **47**, 102742 (2022)

Wang, Y., et al.: A survey on Metaverse: fundamentals, security, and privacy. IEEE Commun. Surv. Tut. **25**, 319–352 (2022)

Wiederhold, B.K., Riva, G.: Metaverse creates new opportunities in healthcare. Annu. Rev. Cybertherapy Telemed **20**, 3–7 (2022)

Yung, K.L., Tang, Y.M., Ip, W.H., Kuo, W.T.: A systematic review of product design for space instrument innovation, reliability, and manufacturing. Machines **9**(10), 244 (2021)

Zallio, M., Clarkson, P.J.: Designing the Metaverse: a study on inclusion, diversity, equity, accessibility and safety for digital immersive environments. Telematics Inform. **75**, 101909 (2022)

Zhang, L., Anjum, M.A., Wang, Y.: The impact of trust-building mechanisms on purchase intention towards Metaverse shopping: the moderating role of age. Int. J. Hum. Comput. Interact., 1–19 (2023). https://doi.org/10.1080/10447318.2023.2184594

Zhang, S., et al.: Towards green Metaverse networking: technologies, advancements and future directions. IEEE Netw. (2023). https://doi.org/10.1109/MNET.130.2200510

Zhang, T., Shen, J., Lai, C.F., Ji, S., Ren, Y.: Multi-server assisted data sharing supporting secure deduplication for metaverse healthcare systems. Futur. Gener. Comput. Syst. **140**, 299–310 (2023)

Zhao, Y., et al.: Metaverse: perspectives from graphics, interactions and visualization. Vis. Inf. **6**(1), 56–67 (2022)

Exploring Research Trends of Metaverse: A Bibliometric Analysis

Sanaa Hassan Zubon Al-Enzi[1], Sammar Abbas[2(✉)],
Abdulnasser AbdulJabbar Abbood[3], Yousif Raad Muhsen[4,5],
Ahmed Abbas Jasim Al-Hchaimi[6,7], and Zainab Almosawi[8]

[1] Technical Institute Basrah, Southern Technical University, Basrah, Iraq
sanaahassanalenzi@stu.edu.iq
[2] Institute of Business Studies, Kohat University of Science and Technology, Kohat, Pakistan
sabbas@kust.edu.pk
[3] Technical Institute Basrah, Southern Technical University, Thi-Qar, Iraq
Abdulnasser.abbood@stu.edu.iq
[4] Department of Computer Science, Faculty of Computer Science and Information Technology,
Universiti Putra Malaysia, Seri Kembangan, Malaysia
yousif@uowasit.edu.iq
[5] Civil Department, College of Engineering, Wasit University, Kut, Iraq
[6] Department of Computer Science, Faculty of Computer Science and Information Technology,
Universiti Putra Malaysia, Seri Kembangan, Malaysia
ahmed.alhchaimi@stu.edu.iq
[7] Southern Technical University, Basra, Iraq
[8] Management Technical College, Southern Technical University, Basrah, Iraq

Abstract. This study aims to determine the development of metaverse research trends published by leading journals on virtual reality. The proliferation of academic and scientific publications regarding virtual reality and virtual networks has increased exponentially. Keeping up with research publications related to the metaverse has become complicated. To this end, in order to provide a summary of the findings of the previous literature and to reorganize the previous literature on the metaverse a bibliometric analysis approach was used. Through the use of VOS Viewer and R-studio, the metaverse-related literature was identified and a comprehensive scientific map of the results of the previous literature was presented. Several results were explored by identifying relevant research and authors, identifying the most popular keywords, and also identifying literary gaps that are considered avenues for future research.

Keywords: Metaverse · Bibliometric analysis · Review · Digital transformation

1 Introduction

Digital transformation has become a usual phenomenon. Businesses are investing millions of dollars for this purpose. According to a recent survey by Boston Consulting Group (BCG), 60% of companies intend to increase their investment in digital technologies. It has been reported that the metaverse has a greater potential to impact digital

M. Al-Emran et al. (Eds.): IMDC-IST 2024, LNNS 895, pp. 21–34, 2023.
https://doi.org/10.1007/978-3-031-51716-7_2

transformation. Riva and Wiederhold (2022) stated that in the next five years, major technology companies will ponder trillions of dollars in investment in the metaverse. The concept of metaverse was first narrated in 1992 by an American author' Neil Stephenson' in his renowned novel 'Snow Crash'. He described the metaverse as a three-dimensional digital world that allows users to escape from the physical world that has become uninteresting to them. He called it a 'shared virtual reality'. Virtual reality (VR) and augmented reality (AR) are considered to be the two technologies at the heart of the metaverse. VR is a computer-generated environment with scenes and objects that appear to be real and make users feel that they are immersed in that environment. For example, flight simulators are used to train pilots. AR is the real-time use of information (text, photos, graphics, etc.) integrated with the real world. It allows digital images and information to be displayed in the physical environment. However, the concept of metaverse as described by Stephenson and the one on which technology companies are working are quite different in a way that the new concept is based on 'inter-reality' i.e., the infusion between real and virtual worlds (Riva et al., 2010; Atiyah and Zaidan, 2022). The concept of the metaverse is built on the belief that our experiences in the physical world influence our virtual world experiences and vice versa. The characteristic feature of the metaverse is that physical objects are connected to the virtual world and living in the virtual world; people are connected to their physical environment (Muhsen et al., 2023; Al-Hchaimi et al., 2023; Chew et al., 2023).

Though the concept of the metaverse is sufficiently older, it started to capture attention in 2020 after the rebranding of Facebook into Meta (Kanterman & Naidu, 2021), which indicated a shift in engagement patterns in the digital world (Suh & Prophet, 2018; Manhal et al., 2023). It has been agreed that the metaverse facilitates the transformation across industries to enrich users' experience through simultaneous engagement in social activities (online meetings, online games) and learning in a virtual environment (Hwang & Chien, 2022). Some may argue that the metaverse is a relatively new term for digital technologies; however, it must be learned that it is beyond the existing technologies (Park & Kim, 2022). It must be considered different from other conventional terms because it offers values like "shared", "persistent," and "decentralized" (Hwang & Chien, 2022: 2). There is little agreement as to the fact that what technologies constitute the metaverse and how the meanings of the metaverse have evolved over the period (Abbas et al., 2023; Bozanic et al., 2023).

In recent years, the bibliometric analysis technique has gained great popularity in the domain of business studies (Donthu et al., 2021; Khan et al., 2021; Gatea and Marina, 2016). This has happened due to easier access and availability of different software such as Gephi, VOS Viewer, and also the large enough databases for example, Scopus and Web of Science. Bibliometric analysis is equally applicable to both information sciences and business studies. Bibliometric analysis can efficiently handle larger scientific data and can produce a higher research impact. Bibliometric analysis is useful for research purposes due to the numerous benefits it offers to researchers. For example, it helps to track the new emerging research areas, patterns of collaborative research, different potential research avenues, and the intellectual developments in a specific area of study (Donthu et al., 2021; Verma & Gustafsson, 2020; Atiyah, 2023); bibliometric analysis has the capacity to handle the large volume of objective data (the number of publications, citations, keywords, topics, etc.). However, the data can be interpreted both objectively and subjectively by doing performance analysis and hematic analysis, respectively. While doing the performance analysis and the thematic analysis, researchers apply established techniques and procedures. Performance analysis determines the research contributions in a specific domain (Cobo et al., 2011; Ramos-Rodrígue & Ruíz-Navarro, 2004; Atiyah, 2023). Performance analysis provides a description of such contribution, which is the promise of bibliometric analysis. Most of the reviews are based on performance analysis as the reviews offer the performance of different research studies in a certain domain and also the performance of the authors, institutions, countries, and journals. It can be said that bibliometric analysis helps to accumulate scientific knowledge and can make sense of emerging fields by making the voluminous unstructured data meaningful. Hence, bibliometric analysis can provide a sound foundation for advancing knowledge in a novel and meaningful way. It assists researchers in attaining a deeper understanding of emerging fields, identifying the research gaps, introducing newer ideas for further investigation, and locating their expected contributions.

Though many studies have used bibliometric analysis for conducting literature reviews, these studies lack a conclusive argument about the effectiveness of biblio-metric analysis. Earlier studies have applied a single or only a few techniques, whereas we have captured more techniques within the bibliometric to comprehend our findings. It is a relatively new technique, and its application across different domains does not fully utilize its potential. This happens because studies are based on limited data, which does not provide a complete and conclusive understanding of the emerging trends in a certain domain. There is the absence of any authentic guidance to use bibliometrics, and this poses a greater challenge to researchers. Keeping in view the mentioned gaps, this study offers a comprehensive use of bibliometrics and accumulates the various techniques of bibliometrics.

2 Methodology

In order to achieve our research objective, we undertook a methodical and comprehensive search. As previously stated, the conceptual advancements and massive influx of fresh data in the metaverse require the implementation of a bibliometric study. In contrast to alternative methodologies, bibliometrics offers a valuable means of examining a substantial corpus of knowledge. It enables the identification of changes within academic fields, the inference of temporal trends, the provision of a comprehensive overview of virtual reality methodologies, and the exploration of advantageous consequences for both scholars and professionals. This study utilizes numerous high-impact journals to adhere to the protocols of systematic search.

To locate the desired publications, researchers conducted a search on Scopus. The authors comprehensively survey the scholarly investigations conducted on the metaverse and its multifaceted applications across several academic domains. The research selection approach encompassed a comprehensive search of pertinent literature, which was contingent upon two successive iterations. Initially, a thorough examination was conducted of the titles and abstracts of the papers in order to exclude any publications that were not relevant or duplicated. Furthermore, following a thorough examination of the complete texts of the selected papers in the initial iteration, a filtration process was conducted. Subsequently, two specialists well-versed in the domain of metaverse were enlisted to verify the screening procedures.

Colleague experts specializing in the domain of metaverse were consulted in order to ascertain probable keywords related to metaverse. Furthermore, based on the keywords identified in earlier investigations within the realm of the metaverse, the prevailing terms (virtual reality) were ascertained. As a result, the subsequent search phrases were employed during the month of May in the year 2023. The query ("metaverse" AND "virtual reality") was utilized in order to acquire the pertinent articles.

All papers that met the specified criteria have been included in the analysis. The primary objective of our study was to categorize Metaverse studies into comprehensive and overarching classifications while also gaining an understanding of the current research trends in this rapidly evolving domain. Following the elimination of duplicate articles, the remaining scholarly works underwent two further rounds of scrutiny and filtration to ascertain their adherence to the requisite criteria for inclusion. 1. The subsequent points outline the justifications for disqualification: The articles are composed in languages distinct from the English language. 2. Any papers that do not primarily focus on the application of the metaverse as a method.

3   Results

In the midst of the copious amount of applied and theoretical literature, staying current with the most recent advancements can present a formidable undertaking. Academic researchers have developed systematic reviews and meta-analyses as methodologies to effectively structure existing research, identify significant concerns, and highlight areas where further investigation is needed (Muhsen et al. 2023). Systematic reviews are of paramount importance in advancing comprehension, refining research endeavors, and consolidating the findings of previous inquiries. Nevertheless, a significant issue associated with systematic reviews pertains to the potential for bias and subjectivity that may emerge when authors are relied upon to interpret and organize the findings of existing literature.

In order to tackle this matter and foster transparency in the dissemination of research findings, certain scholars have suggested employing a comprehensive scholarly research methodology that incorporates the utilisation of the R-tool and VOSviewer. By utilising bibliometrics, this approach prioritises the principles of robustness and transparency in deriving findings from scholarly investigations. The utilisation of R-tool and VOSviewer offers several benefits, encompassing their user-friendly nature and extensive availability, owing to their open-source development and public dissemination. The subsequent sections of this study employ a bibliometric approach to comprehensively examine the subject matter.

Table 1 provides essential details regarding the studies that have been chosen in relation to the metaverse. The provided information encompasses various aspects such as the authors' identities, the geographical location of the conducted studies, the temporal scope of the research, the specific article under consideration, and the criteria employed for rating purposes. The table presents a comprehensive summary of diverse scholarly articles pertaining to the metaverse that have undergone rigorous examination and analysis.

Table 1. Main Information About Data.

Timespan	2009:2023
Sources (Journals, Books, etc.)	213
Documents	316
Annual Growth Rate %	32.55
Document Average Age	0.934
Average citations per doc	6.766
DOCUMENT CONTENTS	
Keywords Plus (ID)	1750
Author's Keywords (DE)	904
AUTHORS	
Authors	1050
Authors of single-authored docs	57
AUTHORS COLLABORATION	
Single-authored docs	65
Co-Authors per Doc	3.73
International co-authorships %	26.58
DOCUMENT TYPES	
Article	165
article conference paper	2
conference paper	92

The primary advantage of utilizing the primary data, as exemplified in Table 1, resides in its capacity to provide a systematic and concise summary of an extensive corpus of scholarly works pertaining to the metaverse. Several notable benefits can be identified. Policymakers, industry professionals, and other stakeholders can employ the condensed data to make well-informed decisions bolstered by an exhaustive examination of extant research pertaining to the subject matter.

The growth of scholarly interest in the metaverse began in 2009, and subsequently, it has garnered increasing attention, culminating in its zenith in 2022. The notable trajectory observed in recent years signifies an increasing captivation and curiosity surrounding the metaverse, rendering it a subject of extensive investigation and a prominent area of discussion within both academic and wider circles. Figure 1 shows the growth of studies.

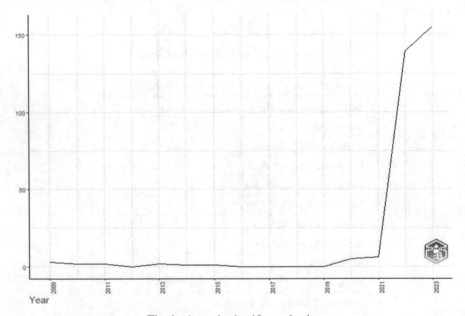

Fig. 1. Annual scientific production.

The increase in interest can be ascribed to multiple factors. The period in question witnessed substantial contributions from technological advancements and innovations, which undoubtedly captivated the interest of researchers, developers, and the wider public (Alnoor et al., 2022). The metaverse, characterized by its virtual environments and immersive encounters, holds the potential to introduce a novel realm of opportunities, fundamentally transforming our modes of interaction, socialization, and commercial activities. The increasing fascination surrounding the metaverse is expected to attract a more significant number of researchers, experts, and investors, thereby facilitating additional progress, innovations, and comprehension within this captivating and ever-evolving domain. The metaverse's popularity and prominence as a subject of active exploration serve as a positive indication of its significance as a transformative and influential concept that has the potential to shape future human interactions and experiences.

Co-occurrence networks are created by analyzing the repeated appearance of terms in the existing academic literature. These networks possess considerable potential to provide valuable insights into the fundamental theoretical frameworks within a particular academic field, thus offering advantages to scholars, researchers, and practitioners. Figure 2 visually depicts the co-occurrence networks, enabling a thorough understanding of the interconnections among frequently utilized terminologies.

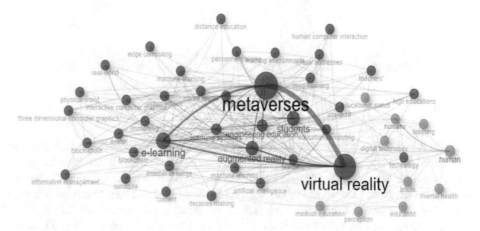

Fig. 2. Co-Occurrence network.

The complex configuration of the co-occurrence network exemplifies the interrelatedness of thematic components identified in the examined literature. Nodes and edges within the network represent significant terms and their respective connections. Significantly, it is apparent that the metaverse holds a prominent position in the discourse of preceding academics when discussing virtual reality. This observation suggests a significant dependence on data networks for the organization of information and interpretation of results within this particular research field.

The scientific collaboration map illustrates the interconnections among various institutions, nations, and authors within the academic community. Collaborative efforts involving multiple writers on a project yield mutual benefits through the collective utilization of each individual's knowledge and expertise in the specific subject matter being developed. Scientific collaboration plays a vital role in promoting the development of academic and industrial institutions. Figure 3 presents a graphical representation of the collaborative efforts within the international metaverse.

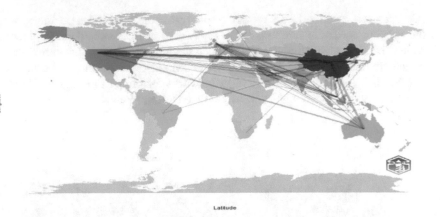

Fig. 3. Country scientific production.

The figure exhibits three discernible colors. The deepest shade of blue indicates the scientific entities with the highest productivity. The presence of a vibrant blue hue indicates a limited level of scientific productivity. The presence of a grey area demonstrates the lack of scientific output. Moreover, the representation of international scientific collaboration is denoted by red lines. To enhance comprehension, researchers hailing from Europe, Australia, and Asia have collaborated in a joint effort (refer to Fig. 3). However, there is a significant deficiency in scientific collaboration between Africa and the Americas. Moreover, the dearth of global scientific collaboration serves as an indication of insufficient acquaintance with and comprehension of the metaverse. Therefore, it is imperative for researchers and policymakers in Africa and the Americas to explore more sophisticated strategies that foster and enhance scientific collaboration, thereby leveraging the knowledge and expertise of European and Australian counterparts in this domain.

The identification of highly influential authors within a specific field is of utmost significance, as it not only serves as a source of inspiration for emerging researchers but also provides them with valuable opportunities for mentorship and guidance in their scholarly pursuits. Moreover, gaining an understanding of author collaboration patterns within the literature can provide valuable insights into the trajectory and future direction of the discipline. Figure 4 provides a succinct overview of the prominent authors in the field, highlighting the significant contributions of Cali and Wang, who are identified as the primary influential figures in this domain. These individuals' noteworthy influence and prominence highlight their substantial contributions to the advancement of the field, emphasizing their potential as valuable sources of knowledge and mentorship for aspiring researchers.

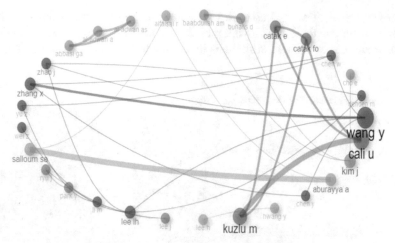

Fig. 4. Co-Occurrence Networks Authors.

This word cloud examines the most commonly utilized and consequential terms extracted from the titles of prior research studies. Figure 5 presents a compilation of essential terminology extracted from the scholarly literature, serving the purpose of providing a concise overview and restructuring the available data. The phenomenon of varying word sizes is observable in multiple contexts.

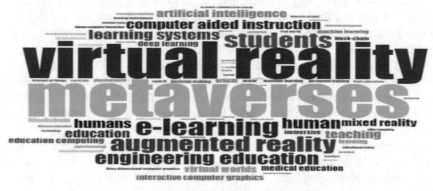

Fig. 5. Cloud of words.

The studies demonstrate that higher word sizes are associated with increased occurrence rates. Terms that have a lower frequency of occurrence in the existing body of literature tend to have less depth or substance. Virtual reality, metaverses, and e-learning are integral components within the current corpus of knowledge in this domain. The literature suggests that the optimization of e-learning with metaverse is a crucial factor to consider in order to enhance virtual techniques.

Journals, universities, and countries all have potentially relevant links. In Fig. 6, we propose a unique three-field layout that depicts the relationships between the most prominent journals on the left, affiliations in the middle, and nations on the right.

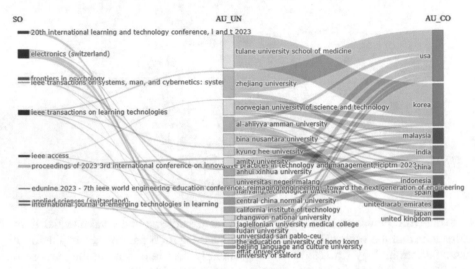

Fig. 6. Three-Field Plot.

We discovered that the majority of papers on this issue were produced by writers from the United States and Koria and were published in electronic journals and IEEE. The majority of academic affiliations in this field are with Tulane University and Zhejiang University. Overall, Fig. 8 gives a positive image on which new authors can rely.

To illuminate the progression of this particular field of study, it is imperative to ascertain the scholarly works that have exerted a substantial influence on the existing body of literature. Similarly, gaining knowledge about the patterns of citations within the existing literature could offer valuable insights into the prospective trajectory of the field. Table 2 provides an explanation of the most significant studies.

Table 2 provides a comprehensive summary of the ten most influential papers that have significantly impacted the field. The three most notable papers were published in the following academic journals: 1- International Journal of Information Management, 2- IEEE Access, and 3- Clinical Health. A significant proportion of the most crucial and influential scholarly articles provide both theoretical and practical contributions. Theoretical issues and their corresponding resolutions are frequently expounded upon and offered within these scholarly contributions.

Table 2. Most important studies.

Journal	DOI	TC
International Journal of Information Management	https://doi.org/10.1016/j.ijinfomgt.2022.102542	185
IEEE Access	https://doi.org/10.1109/ACCESS.2022.3169285	61
Clinical Health	https://doi.org/10.1016/j.ceh.2022.02.001	43
Sustainability (Switzerland)	https://doi.org/10.3390/su14084786	37
Frontiers In Psychology	https://doi.org/10.3389/fpsyg.2022.1016300	14
Computers and Education	https://doi.org/10.1016/j.compedu.2022.104693	7
Frontiers in Psychology	https://doi.org/10.3389/fpsyg.2022.1016300	14
Southeastern European Journal of Public Health	https://doi.org/10.11576/seejph-5759	9
Proceedings - 2022 IEEE 42nd International Conference on Distributed Computing Systems Workshops, ICDCSW 2022	https://doi.org/10.1109/ICDCSW56584.2022.00053	6

4   Why the Bibliometric

The results of this bibliometric analysis have significant implications for assessing and comprehending research outcomes pertaining to the metaverse. Several noteworthy implications can be identified: The process of identifying key contributors involves conducting an analysis to determine the authors and researchers who have had the greatest impact on the field of the metaverse. The identification of these significant contributors can offer valuable insights into the expertise and knowledge foundation that underpins progress in the field. In addition, the study provides an assessment of research trends and areas of focus within the metaverse through the analysis of co-occurrence networks and collaboration patterns. This aids in comprehending the dynamic nature of the discipline and its trajectory. Moreover, the evaluation of research impact can be facilitated through bibliometric analysis, which allows for the assessment of the influence and significance of individual studies and research papers. Through the examination of citation counts and co-citation networks, scholars are able to assess the importance and impact of specific scholarly works within the wider academic community. Additionally, the identification of research gaps involves the examination of areas within metaverse research that have received limited attention or remain underexplored. The process of identifying research gaps can serve as a valuable tool in guiding future investigations and determining the priority of research topics that require further exploration. Furthermore, the bibliometric findings can serve as a valuable tool for policymakers and funding agencies, enabling them to make well-informed decisions about allocating resources and developing policies in domains associated with the metaverse. This facilitates the reinforcement of research endeavors that are in accordance with the present state of the discipline and its potential influence on diverse sectors. The promotion of interdisciplinary collaboration can be facilitated through the utilization of co-occurrence networks, which have the ability to identify and emphasize potential connections and collaborations among researchers who are engaged in metaverse-related studies. The cultivation of interdisciplinary collaborations has the potential to yield novel methodologies and comprehensive resolutions to intricate problems. Next, improving the Processes of Literature Review: The analysis offers a systematic and thorough examination of the current body of literature pertaining to the metaverse, facilitating researchers in conducting literature reviews that are both efficient and impactful. This facilitates the process of obtaining pertinent information and enhances the groundwork for novel research undertakings.

5 Discussion and Conclusion

Bibliometric analysis is an advanced technique for pursuing and accumulating scientific knowledge in well-established and emerging areas of business research. It is equally useful for both seasoned and early-career researchers who want to surface novel research ideas. In recent times, bibliometric analysis has gained attention and popularity due to the availability of a variety of software that can assist in performing comprehensive literature reviews. Also, the availability and access to the different databases help researchers to manage the volume of raw and unstructured data. Bibliometrics is not only useful for business research but also equally useful for studies in the domains of artificial intelligence and big data (Makarius et al., 2020; Mustak et al., 2021; Alnoor et al., 2022). It has been suggested that the techniques of bibliometric analysis used to perform the reviews are critical and equally important. The choice of techniques significantly influences the results and the conclusion to be drawn from the performance and subjective analysis. It is quite important to realize that bibliometric analysis is an effective tool for summarizing and synthesizing literature; however, it has some limitations as well. For example, the data obtained from the databases is not exclusively for performing the bibliometric analysis and also contains errors that can influence the analysis to be performed at later stages using such erroneous data. In order to avoid such errors in the data, the researchers must carefully filter the data by removing duplicate data. Similarly, the two streams of the analysis (quantitative vs. qualitative) can explain the underlying relationships, which may be quite unclear occasionally. (Wallin, 2005). In that case, the researcher must take help from other analytic tools, such as content analysis, to better and clearly narrate the relationship in the data. Bibliometric analysis can offer only shorter-term predictions about emerging areas, and hence, the researchers do not need to be ambitious while drawing conclusions about future research directions (Gaur & Kumar, 2018; Albahri et al., 2022).

Despite its limitations, bibliometric analysis encourages early-career researchers to avoid the fear of handling a large amount of data and to pursue their research goals. Bibliometric analysis enhances our understanding of existing knowledge and facilitates knowledge creation for the development of future research agendas. Bibliometric analysis enhances our understanding of previous literature and increases the trustworthiness of the accumulated knowledge. Bibliometric analysis techniques are statistical tools and allow for results production by researchers working independently (Cram et al., 2020; Peng, 2011; Abdullah et al., 2022). These techniques are applicable for confirmatory and exploratory studies and provide objective explanations of the knowledge produced. The major advantage of bibliometric analysis lies in its feature to identify the sub-fields in a major domain.

References

Abbas, S., Alnoor, A., Yin, T.S., Sadaa, A.M., Muhsen, Y.R., Khaw, K.W., Ganesan, Y.: Antecedents of trustworthiness of social commerce platforms: a case of rural communities using multi group SEM & MCDM methods. Electron. Commer. Res. Appl. **62**, 101322 (2023)

Abdullah, H.O., Atshan, N., Al-Abrrow, H., Alnoor, A., Valeri, M., Erkol Bayram, G.: Leadership styles and sustainable organizational energy in family business: modeling non-compensatory and nonlinear relationships. Journal of Family Business Management **13**, 1104–1131 (2022)

Albahri, O.S., et al.: Novel dynamic fuzzy decision-making framework for COVID-19 vaccine dose recipients. J. Adv. Res. **37**, 147–168 (2022)

Al-Hchaimi, A.A.J., Sulaiman, N.B., Mustafa, M.A.B., Mohtar, M.N.B., Hassan, S.L.B.M., Muhsen, Y.R.: A comprehensive evaluation approach for efficient countermeasure techniques against timing side-channel attack on MPSoC-based IoT using multi-criteria decision-making methods. Egyptian Inform. J. **24**(2), 351–364 (2023)

Alnoor, A., Khaw, K.W., Al-Abrrow, H., Alharbi, R.K.: The hybrid strategy on the basis of miles and snow and porter's strategies: an overview of the current state-of-the-art of research. Int. J. Eng. Bus. Manage. **14**, 18479790221080216 (2022)

Alnoor, A., et al.: Toward a sustainable transportation industry: oil company benchmarking based on the extension of linear diophantine fuzzy rough sets and multicriteria decision-making methods. IEEE Trans. Fuzzy Syst. **31**(2), 449–459 (2022)

Atiyah, A.G.: Power distance and strategic decision implementation: exploring the moderative influence of organizational context. Int. Acad. J. Bus. Manage. **10**, 71–80 (2023)

Atiyah, A.G.: Strategic network and psychological contract breach: the mediating effect of role ambiguity. Int. J. Res. Manage. Stud. **13** (2023)

Atiyah, A.G., Zaidan, R.A.: Barriers to using social commerce. In: Alnoor, A., Wah, K.K., Hassan, A. (eds.) Artificial Neural Networks and Structural Equation Modeling. Springer, Singapore (2022). https://doi.org/10.1007/978-981-19-6509-8_7

Bozanic, D., Tešić, D., Puška, A., Štilić, A., Muhsen, Y.R.: Ranking challenges, risks and threats using fuzzy inference system. Decis. Mak. Appl. Manage. Eng. **6**(2), 933–947 (2023)

Chew, X., Khaw, K.W., Alnoor, A., Ferasso, M., Al Halbusi, H., Muhsen, Y.R.: Circular economy of medical waste: novel intelligent medical waste management framework based on extension linear diophantine fuzzy FDOSM and neural network approach. Environ. Sci. Pollut. Res. Int. **30**(21), 60473–60499 (2023)

Cobo, M.J., López-Herrera, A.G., Herrera-Viedma, E., Herrera, F.: An approach for detecting, quantifying, and visualizing the evolution of a research field: a practical application to the fuzzy sets theory field. J. Informet. **5**(1), 146–166 (2011)

Cram, W.A., Templier, M., Paré, G.: (Re) considering the concept of literature review reproducibility. J. Assoc. Inf. Syst. **21**(5), 10 (2020)

Donthu, N., Kumar, S., Pandey, N., Lim, W.M.: Research constituents, intellectual structure, and collaboration patterns in journal of international marketing: an analytical retrospective. J. Int. Mark. **29**(2), 1–25 (2021)

Gatea, A.A., Marina, V.: Higher education funding in Iraq in terms of the experience of particular developed countries. Int. J. Adv. Stud. **6**(1), 8–17 (2016)

Gaur, A., Kumar, M.: A systematic approach to conducting review studies: an assessment of content analysis in 25 years of IB research. J. World Bus. **53**(2), 280–289 (2018)

Hwang, G.J., Chien, S.Y.: Definition, roles, and potential research issues of the metaverse in education: an artificial intelligence perspective. Comput. Educ. Artif. Intell. **3**, 100082 (2022)

Kanterman, M., Naidu, N.: Metaverse may be the $800 billion market, next tech platform. Bloomberg Intell. (2021)

Khan, M.A., Pattnaik, D., Ashraf, R., Ali, I., Kumar, S., Donthu, N.: Value of special issues in the journal of business research: a bibliometric analysis. J. Bus. Res. **125**, 295–313 (2021)

Makarius, E.E., Mukherjee, D., Fox, J.D., Fox, A.K.: Rising with the machines: a sociotechnical framework for bringing artificial intelligence into the organization. J. Bus. Res. **120**, 262–273 (2020)

Manhal, M., Al-khalidi, A., Hamad, Z.: Strategic network: managerial myopia point of view. Manage. Sci. Lett. **13**(3), 211–218 (2023)

Muhsen, Y.R., Husin, N.A., Zolkepli, M.B., Manshor, N., Al-Hchaimi, A.A.J.: Evaluation of the routing algorithms for NoC-based MPSoC: a fuzzy multi-criteria decision-making approach. IEEE Access **11**, 102806–102827 (2023). https://doi.org/10.1109/ACCESS.2023.3310246

Mustak, M., Salminen, J., Plé, L., Wirtz, J.: Artificial intelligence in marketing: topic modeling, scientometric analysis, and research agenda. J. Bus. Res. **124**, 389–404 (2021)

Park, S.M., Kim, Y.G.: A metaverse: taxonomy, components, applications, and open challenges. IEEE access **10**, 4209–4251 (2022)

Peng, R.D.: Reproducible research in computational science. Science **334**(6060), 1226–1227 (2011)

Ramos-Rodríguez, A.R., Ruíz-Navarro, J.: Changes in the intellectual structure of strategic management research: a bibliometric study of the Strategic management journal, 1980–2000. Strateg. Manag. J. **25**(10), 981–1004 (2004)

Riva, G., Wiederhold, B.K.: What the metaverse is (really) and why we need to know about it. Cyberpsychol. Behav. Soc. Netw. **25**(6), 355–359 (2022)

Riva, G., et al.: Interreality in practice: bridging virtual and real worlds in the treatment of posttraumatic stress disorders. Cyberpsychol. Behav. Soc. Netw. **13**(1), 55–65 (2010)

Suh, A., Prophet, J.: The state of immersive technology research: a literature analysis. Comput. Hum. Behav. **86**, 77–90 (2018)

Verma, S., Gustafsson, A.: Investigating the emerging COVID-19 research trends in the field of business and management: a bibliometric analysis approach. J. Bus. Res. **118**, 253–261 (2020)

Wallin, J.A.: Bibliometric methods: pitfalls and possibilities. Basic Clin. Pharmacol. Toxicol. **97**(5), 261–275 (2005)

Inducing Virtual Reality to Improve Human-Virtual Social Interaction and Learning Perceptions

Ahmad Al Yakin[1]([✉]), Luís Cardoso[2], Abdul Latief[1], Muthmainnah[3], Muhammad Arsyam[4], and M. Yusri[4]

[1] Teacher Training and Education Faculty, Pancasila and Civic Education Study Program, Universitas Al Asyariah Mandar, Polwali Mandar, Indonesia
ahmadalyakin76@gmail.com, latief2002@gmail.com

[2] Polytechnic Institute of Portalegre and Centre for Comparative Studies, University of Lisbon, Lisbon, Portugal
lmcardoso@ipportalegre.pt

[3] Teacher Training and Education Faculty, Indonesian Language Study Program, Universitas Al Asyariah Mandar, Polewali Mandar, Indonesia
muthmainnahunasman@gmail.com

[4] Tarbiyah, Sekolah Tinggi Agama Islam Darud Da'wah Wal-Irsyad Makassar, Makassar, Indonesia
arsyam0505@gmail.com, myusrirk@gmail.com

Abstract. The aim of this study is to investigate Human-Virtual social interaction in a metaverse learning environment and draw conclusions about the efficacy of metaverse technologies in education, namely in (target segment). The main research was conducted by analysing survey data and written comments from respondents. This research demonstrates that the human-virtual social collaboration that occurs during immersion performed using the metaverse is significantly predicted by the use of teacher pedagogy and the influence of this aspect of technology. The results support the findings of previous studies, which highlight the importance of technology-enabled learning environments when teaching using virtual reality (VR) technologies. This study also shows through the recorded statistical data in the form of feedback and observations that learning is more interesting and increases student engagement by utilizing social interaction through the Metaverse classroom learning model. This research also reveals that social interaction with the metaverse learning environment helps to create increased human interaction i.e. with classmates and teachers, and is accompanied by increased interaction with machines simultaneously, making it a valuable tool for teachers in aligning the Merdeka curriculum in Indonesia. In conclusion, the overall results prove that metaverse technology can encourage active participation, engagement, enthusiasm, and achievement of learning goals in the classroom.

Keywords: Metaverse · Virtual reality · Human-Virtual interaction · Social Interaction and social education

M. Al-Emran et al. (Eds.): IMDC-IST 2024, LNNS 895, pp. 35–48, 2023.
https://doi.org/10.1007/978-3-031-51716-7_3

1 Introduction

The trend of commercialization of education and research can be revolutionized by the metaverse. In the past, a true lecturer—a rare commodity in universities—would speak physically in front of a select group of students. When the virtual and real worlds intersect in the metaverse, it will provide students with a more "cyber-physical" learning environment. In the cyber-physical learning environment students can easily switch between online learning and a hybrid classroom while using the same avatar (Chua & Yu, 2023; Awotunde et al., 2023; Hamid et al., 2021). The metaverse will enhance these changes because conventional tertiary education is also being transformed into modern education. Being exposed to cyberphysics has the potential to attract large numbers of university students and accelerate their learning more efficiently (Asad & Malik, 2023; Eneizan et al., 2019). In the metaverse, they may take advantage of virtual experiences provided by educational institutions around the world and can easily adapt to these technologies (Muhsen et al., 2023; Al-Hchaimi et al., 2023; Chew et al., 2023).

Researchers and programmers may work together soon to design and build trainers that can assist educators in metaverse-based environments (Koohang et al., 2023; Al-Abrrow et al., 2021). Due to its recent creation, the metaverse system is being studied by researchers interested in its potential impact on education more broadly. The Metaverse System is a new approach to online or hybrid education problems that has been offered as a means to solve them. The challenges that teachers face when transferring certain classes and the extent to which students enjoy online learning are the two most prominent problems that can be overcome (Hodges, 2023; Khaw et al., 2022). In this digital era, a learning environment with a metaverse approach is believed to facilitate the widespread use of learner-centered active learning models such as simulations and game-based learning (Han et al., 2023; Alsalem et al., 2022; Fadhil et al., 2021). Without worrying about the real-world repercussions of their actions, students will be able to experiment with equipment and hone complex procedural and behavioural skills in the Metaverse (Yilmaz et al., 2023; Albahri et al., 2021). It is thought that the Metaverse, with the help of technologies such as the Internet of Things, Artificial Intelligence, and machine learning, can enhance students' ability to engage with physical and digital environments (Ning et al., 2023; Alnoor et al., 2022). However, the educational potential of the Metaverse has not been thoroughly investigated in the social education literature (Alam & Mohanty, 2022; AL-Fatlawey et al., 2021). Therefore, it is important to study Human-Computer Interaction (HCI) preferences for potential future hybrid classroom learning environments based on the Metaverse approach.

The technical rebirth of the Metaverse can inspire new interaction metaphors, ideas, and tools, which in turn can shape how people engage with this cutting-edge platform (Abbas et al., 2023; Bozanic et al., 2023). From this HCI perspective, although much has been written about the use of wearable technology in VR games research on the Metaverse platform, particularly outside the gaming industry i.e. education is still lagging. This goes against the belief that the Metaverse will soon become the next ubiquitous computing paradigm, which has the power to revolutionize many aspects of e-business such as distance learning or online and hybrid education (Perisic et al., 2023; Gatea, and Marina, 2016). Current Metaverse HCI research focuses on more intuitive methods of human-computer interaction, such as talking to machines or using expressions and gestures. Several early studies looked at the feasibility of using eye path tracking instead of traditional computer input peripherals such as a key-board and mouse in the game's Metaverse environments (Zhang, 2023; Manhal et al., 2023). When compared to traditional input methods such as clicking and moving the mouse, view control proves to be much more enlightening. The relevance of geo-graphic location mapping in immersive platforms for creating a sense of reality and maximizing the presence of "being there" was highlighted through the use of eye tracking to evaluate individualized cognitive training built for realistic route learning (Atiyah, and Zaidan, 2022).

Research on user preferences for navigation on virtual platforms, for example, could be the forerunner of future Metaverse-Learning HCI studies. Several VR-related studies highlight elements that influence the user experience with VR drivers, such as psychophysical discomfort, immersion, ease of use, familiarity, competence, and efficacy. A related study, evaluated three different actuation methods for use in a VR setting to see how they affect the user experience. While navigating VR settings, users have mentioned issues such as effectiveness, motion sickness, presence, and personal preference after experimenting with VR (Boletsis & Cedergren, 2019; Atiyah, 2023). The social constructivist view is concerned with how factors such as culture, social interaction, classroom settings, and individual prior knowledge shape their learning and subsequent knowledge production. Learning, according to Vygotsky (Lee, 2014; Atiyah, 2020), "evokes various internal developmental processes capable of operating only when [the learner] interacts with people in his environment and cooperates with his peers" or builds a solid work team. Because knowledge is rooted in the social environment, it is first co-constructed or co-created with others at the inter-psychological level, or, in other words, children acquire the societal awareness necessary to master a concept and then the individual awareness necessary to apply that idea to new situations such as the Metaverse environment.

The relationship between the Social Constructivist theory and the HCI theory is that they have similarities in emphasizing the importance of creative activity in the eductional process. The ability to generate new or authentic ideas and visible products that can carry scientific, aesthetic, social, or technological value in specific socio-cultural and environmental contexts of use is developed through this process. The Zone of Proximal Development, as defined by Vygotsky in 1978, is the area where a learner's actual development meets his or her potential. The teacher's responsibility is to fill this gap by providing students with the tools, media, or technology and support they need to succeed. The goal of education is to help students develop their meaning through collaborative problem-solving with classmates, teachers, and the surrounding learning environment. It is important to investigate the social interaction that is elicited, even though it does not provide a complete explanation of how experience, understanding, and knowledge are acquired and modified from this social interaction (Zhu & Zhang, 2023). The capabilities inherent in VR, such as scaffolding and the creation process during interaction, can help students gain maximum interaction skills in the class-room.

As we know, several studies have focused on the adoption and acceptance of metaverse systems by users in industrialized countries, as they have recently been designed to solve challenges in online learning settings. Recent metaverse research has shown that users value Metaverse as a means of making classroom management simpler, interesting, and exciting (Al-Adwan et al., 2023). As a result, the findings of this study have significant implications for the adoption and use of the metaverse of technology in terms of eliciting social interaction. The importance of the metaverse has been the subject of several studies, but none of these studies have provided satisfactory results due to the lack of a comprehensive model covering all the key aspects. One of the main reasons for the lack of a complete understanding of the efficiency of the Metaverse is that it is a newly developed system that is still rarely used, especially in developing countries, from elementary to university levels. The goal of this investigation is for future researchers to use robust methods and a comprehensive theoretical framework to examine the ramifications of the Metaverse system.

In this research, we target one application of VR technology: human social interaction, where one person communicates with another person (both real and virtual) using the CoSpace Edu platform. As previously explained by (Cao et al., 2023; Chen et al., 2023), the learning environment with the Metaverse approach based on VR has been extensively used in research about spatial cognition, learning performance, and motor control. As higher education educators, we are especially interested in developing VR for use in social scientific studies and extending the benefits of the metaverse approach as teaching material in educational sociology. Please note that the word VR refers to a computer-generated or smartphone-generated world and not just an item visible through a head-mounted screen. The last category incorporates technologies such as 360-degree video while excluding augmented reality and other non-immersive computer-generated systems. This study focuses on VR that is downloaded to a computer or smartphone and used as IT-based teaching materials. The context of this study sees the VR world as uncharted territory, with social interactions acting as explorers to generate human-virtual interactions in the classroom. We liken the obstacles that this explorer will face when conducting research utilizing VR to social phenomena contained in the VR CSpace Edu platform. We start from the bottom, define the important social science themes our explorers need, and outline the upcoming terrain in surveying the social issues that must be tackled when immersed in VR. Second, many people can overcome social problems with the right solutions; similarly, we discuss the potential problems with adopting social VR scenarios and the best results that can be achieved with current technology. Finally, we discuss the biggest problems in creating fully interacting virtual humans and provide proposals for how computers and social theory should come together to achieve this goal. In this paper, we wish to provide a balanced assessment of the state of VR by discussing its promise and limitations for enhancing social interaction.

2 Method

This research follows a quantitative research methodology by surveying the effect of learning with a metaverse approach on students' social interaction skills using VR using a survey questionnaire. Due to the emphasis on objectivity in this research's explanation of the phenomena investigated, it is consistent with a postpositivist worldview (Creswell, 2018).

2.1 Participants

Undergraduate students from Al Asyariah Mandar University who participated in the metaverse learning project were used as the study population of 350 undergraduate students. The participants were randomly selected from six faculties. A total of 87 valid responses were found as a sample from the computer science faculty, namely 48 undergraduate students from the information systems study program and 39 students from the information engineering study program; they consisted of 47 male students and 40 female students respectively.

2.2 Procedure

During the first phase, this metaverse learning approach is implemented by providing a shared interactive display that provides a digital workspace in a computer lab. Social interaction greatly benefits from co-working spaces, which increase awareness of other people's actions and concurrent interactions. This activity is to observe social interactions that take place in the metaverse class environment.

In the second phase, learning activities in this study allow people to collaborate on one computer that is accessed by two or three students. Smartphone use is also permitted to access VR with shared interactive displays. Although the small size of mobile phones naturally limits the input and output capacity of communal use, mobile phones are also used to implement the co-working space concept. As a result, learning models designed to provide each user with their own set of tools are often used to increase the area available for engagement and level the playing field for everyone taking part in a joint effort enabling informal group sharing. In addition, users can use mobile devices and large screens simultaneously to create highly adaptable interaction areas for group work. In short, this strategy is widely used in systems that provide classroom interactions to be observed, namely associative and dissociative processes.

Fig. 1. Students' social interaction activities.

2.3 Questionnaire

Respondents' answers were collected using a 20-item survey questionnaire. To achieve the research objectives, we integrated and modified questionnaire items from (Salloum et al., 2023) and included them in the research questionnaire. The survey uses a five-point Likert Scale.

3 Data Analysis

This research questionnaire uses a Likert scale to measure how the metaverse learning model using VR can influence student interaction patterns during learning if the respondents indicate that they strongly agree with the statements given by the scores of the questionnaire answers. The Likert scale range applied is 0–5 and with this term, the user is forced to take a stand. Choosing a "neutral" option is out of the question. The 5-point scale they use is ideal for academics because it encourages thoughtful responses in the categories of strongly agree (5), agree (4), disagree (3), disagree (2), and strongly disagree (1).

This category is very efficient at gathering feedback from customers about the products and services they have experienced.

Having strong social skills is important for maintaining healthy relationships and working efficiently with other people and communities. Dedication to helping others, working together, making decisions, communicating, caring for the environment, starting a business, and being active in the community are examples of social skills. Students need to learn (1) self-direction (learning independently or autonomously and critically); (2) self-control (cooperation, tolerance, respecting the rights of others to express opinions, and having social sensitivity or empathy for classmates who are below average); and (3) sharing ideas and experiences as well as planning, evaluating, and collaborating with technology, which is a form of social interaction that occurs in the classroom between virtual human beings, which will be described as follows (Table 1).

Table 1. Students questionnaire survey result.

Sample size profile			Statistic	Std. Error
Social interaction	Mean		86.805	0.699
	95% Confidence Interval for Mean	Lower Bound	85.415	
		Upper Bound	88.195	
	5% Trimmed Mean		86.864	
	Median		86.000	
	Variance		42.531	
	Std. Deviation		6.522	
	Minimum		75.000	
	Maximum		99.000	
	Range		24.000	
	Interquartile Range		10.000	
	Skewness		0.163	0.258
	Kurtosis		1.135	0.511

Based on the results of descriptive analysis of the social interaction variables that occur during learning using VR, it is known that the maximum score obtained from the student questionnaire is 99 and the minimum score obtained by students is 75. The average score obtained by students is 86.80, and the standard deviation obtained is 6.52 from the ideal score that students might achieve. While the skewness and kurtosis values are 163 and 1.135, respectively.

Table 2. Normal parameters of the data.

N		VR Social Interaction
		87
Normal Parameters[,b]	Mean	86.8046
	Std. Deviation	6.52159
Most Extreme Differences	Absolute	0.128
	Positive	0.114
	Negative	0.128
Test Statistic		0.128
Asymp. Sig. (2-tailed)		.001[c]

Based on Table 2, Tests of Normality (One-Sample Kolmogorov-Smirnov), the Kolmogorov-Smirnov test value is obtained with a significance of 0.001. The value obtained from the SPSS output results shows that the research data is normally distributed.

Table 3 shows the social interaction of undergraduate students with one another is characterized by a two-way flow of effects: one positive and one negative, due to the receptive and responsive nature of the two parties involved. Communication and social interaction arise when two people connect to achieve a common goal. Thus, social contact can be described as a two-way relationship between individuals. In this activity, undergraduate students discuss material with each other and conclude interactions with VR, namely museums, firefighters, Life of Pi, and cities. They found that through this collaboration, they shared the understanding and knowledge gained from VR applications and related it to reality based on their respective experiences.

Table 3. Mean score and standard deviation in dealing with individual-individual social interaction.

Items	N	Mean	Std. Deviation
1. How comfortable do you feel participating in class discussions or sharing ideas with your colleagues?	87	4.609	0.491
2. How comfortable do you feel participating in class with metaverse model?	87	4.609	0.491
3. I enjoy collaborating with classmates on group projects or assignments	87	4.609	0.491
4. I engage in conversations with my classmates while studying with the metaverse learning model	87	4.241	0.430
5. I feel included and accepted by my classmates during class activities	87	4.655	0.478
6. I seek help or support from classmates when I have difficulty studying	87	4.644	0.482
7. I feel more comfortable interacting with friends regarding material in VR	87	4.874	0.334
8. The existence of VR encourages positive social interaction in the classroom. VR technology provides a challenge so that social interaction with classmates and teachers becomes even more intense	87	4.310	0.556
9. I often engage in informal conversations or small talk with classmates before or after class, discussing the lecturer's VR projects	87	4.310	0.811
10. I believe classmates respect and value the opinions and contributions made during class discussions	87	4.058	0.491

Table 4 shows the human interaction with computers or students and VR. The data from the student survey regarding their interaction with VR is also positive, they feel confident to learn, share ideas, and discuss the topic of the materials from VR. Not only that, but VR also help them to understand the materials more easily and feel comfortable and this technology gives them have great start topic to human-virtual interaction at the same time.

Table 4. Mean score and standard deviation in dealing with human-computer interaction.

Items	N	Mean	Std. Deviation
1. I often use a computer or digital device in my daily life	87	4.368	0.485
2. I feel comfortable when interacting with virtual reality	87	3.724	0.802
3. Have you ever experienced difficulty or frustration while using Virtual reality?	87	4.322	0.690
4. I am very pleased with the presence of virtual reality applications to help understand the subject matter	87	4.149	0.856
5. I believe the responsiveness of the VR interface really helps me express ideas and feedback. The design and layout of the VR feature interface influence my overall learning experience and productivity	87	4.149	0.856
6. I am satisfied with the level of personalization and customization options. VR excites the imagination	87	4.149	0.856
7. I feel that the interface of the VR device has increased my motivation to discuss, share opinions, and think critically	87	3.989	0.581
8. I receive positive feedback or confirmation when interacting with VR	87	4.115	0.672
9. I like to adapt my behavior or learn new skills to use VR effectively	87	4.460	0.501
10. I felt confident to discuss, share ideas, and voice opinions with friends while interacting with this VR	87	4.460	0.501

4 Discussion

In this study, the metaverse approach to learning influences the social interaction patterns of undergraduate students. The observation results show that there is an increase in good cooperation among group mates or teams during learning, and students will have good strategies for learning the right way, practicing language skills, making new friends, developing analytical skills, critical thinking, and increasing social skills in the digital world. In addition, interaction with computers provides awareness of the importance of using technology in learning, not only for playing games or similar entertainment. The presence of a metaverse environment in the hybrid class promises students to be much smarter in technology; VR helps students' motor development, and VR also becomes an encouragement for students to learn. With the emergence of VR, students' social interaction patterns have increased because they are more focused on learning and trying to analyse phenomena in social media and relate them to the real world. This comparative analysis between the real and virtual worlds aroused students' enthusiasm to interact with their friends and lecturers. They enjoy learning VR in class and at home, so learning with this project-based outcome keeps them connected to the machine and their classmates whenever and wherever.

This research has been proven to be an effective and strong educational strategy to become a reference for social interaction. The learning activities involved are students motivating and supporting one another by asking questions, explaining, and justifying ideas, unifying ideas, concluding, evaluating data, then articulating reasoning, and providing opportunities to describe and reflect on knowledge gained from VR applications with experience in the real world (Table 3). The interaction activity designed in this study involves placing students in groups to work well together to learn and provide critical thinking about the phenomena identified in VR applications. During learning, it seems that undergraduate students can achieve a healthy balance of interaction with computers HCI, competition in solving social criticism of phenomena contained in VR, and cooperation, support, understanding, and encouragement if they have difficulty operating VR. Learning with this metaverse model has a better value and maybe even an overall educational value when all undergraduate students are actively involved in the interaction. Some students participating in team projects or exposed to integrated work environments acquire the social skills necessary to operate effectively in computer-induced (technology) teams (Table 4).

Teachers instill in their students not only knowledge of the subject matter but also the interpersonal and collaborative skills they need to succeed. Students who are engaged in a metaverse environment, such as using VR technology to learn, also need direction and support from the instructor, just as they would in a traditional class-room setting. The effects of this metaverse environment can increase the necessary social interaction skills. HCI plays an important role here and provides opportunities for practice, feedback, and direction to acquire this ability when operating or playing with VR. This research has filled this knowledge gap by getting maximum results. According to the findings of this study, college students studying with group projects and discussions rather than individual study can increase social interaction in the classroom. Most students surveyed feel that the metaverse learning model is beneficial to students and should be used in the classroom (Muthmainnah et al., 2023) we find that mostly the students interact more often such as discussing, sharing ideas or ideas, and asking for opinions and arguments. Based on these findings, the metaverse learning model helps students develop their abilities to learn from and teach one another. Through learning VR, friends can help each other when one student has difficulty operating or encounters problems during learning. The results showed that students' conversation skills improved when they were involved with VR like (Kaddoura & Al Husseiny, 2023) studied. Positive interdependence, promotion of face-to-face interaction, individual accountability, interpersonal or small group skills, and technology processing have all been identified as important components of successful and effective patterns of social interaction. The results of this study indicate that when students work together to solve a problem in a metaverse learning environment, they are more likely to share ideas. Children show genuine curiosity by being patient and empathetic in their interactions with peers from various cultural backgrounds. This interaction is balanced, between human and virtual. This study indicated that previous studies (Kaldarova et al., 2023) have shown that students learn more at ease, are happier, are more motivated (Mitsea et al., 2023) and engaged, and their interactions increase when they are immersed in VR, have better problem-solving (Al Yakin & Seraj, 2023) and communication skills, and ultimately do better academically.

The results of this study wish to provide a unique theoretical contribution that provides a framework to assist in the development of technology to better suit VR users who are physically present in the same room and interact. Therefore, as we will show throughout the findings of this study, this exploration of metaverse use design is based on certain challenges or traits related to social interaction. Social interaction centers on a group of people residing in the same location whose main interactions are with technological artefacts. HCI interactions in this context are not only related to interactions between humans and computers but also between humans. Human-computer interaction often blends with other types of human-to-human and human-to-mediator interactions that technology plays in the highly intertwined social interactions of the metaverse. The social interaction that takes place in this study is social contact and communication that occurs reciprocally between humans and virtual. In the classroom, direct social contact activities occur between students and lecturers (Fig. 1), and direct contact with computers HCI occurs simultaneously with the activity of meeting someone directly and engaging in conversation. Interactions with other people can have both beneficial and detrimental effects. Conversely, positive social interaction such as cooperation is analogous to fighting. The interaction of students with their peers also instructs them in communication, attitude, compassion, and empathy, which can foster character within them by continuously changing their attitude and language. In this way, students' intellectual and emotional well-being can be enhanced using learning materials that encourage social interaction as well as shaping the spectrum of positive interactions undergraduate students will adopt out-side the classroom and in their work environment.

5 Conclusion

Education is known as the best way to acquire social skills. The results of the education obtained are an increase in students' knowledge, abilities, and dispositions. These three features should be practical in everyday situations. For this reason, the metaverse learning model has been successfully implemented in a VR setting, benefitting students in the field of computer science. The findings of this research add to the growing body of evidence supporting the value of social interaction methods derived from instructional media and technology, which seek to build mutually bene-facial dynamics between human instructors, human students, and computers inside and outside the classroom. The VR learning style was well received by undergraduate students, who reported maintaining positive relationships with classmates and professors. Most students agree that working together on projects is more enjoyable and productive than working alone. Students are expected to use the social skills they learn in this course to become productive members of society not only in the class-room but can practice them in everyday life. It is also anticipated that students will develop their social networks, demonstrate social awareness, demonstrate empathy, critical thinking skills, problem-solving, caring, and use democratic ways of solving problems. In summary, this work assists in the analysis of previous research, elucidation of the contributions of new research, and their placement in the landscape for future research in the metaverse field. Future researchers will be able to adapt different metaverse learning models based on the cases they find thanks to the classifications offered here, which help identify meaningful goals and approaches to growing social interaction in the classroom. We hope that our research will encourage others to contribute to the discussion of how traditional pedagogical approaches used so far can be adapted in a modern way for use in metaverse settings accessible via computers or smartphones.

References

Abbas, S., et al.: Antecedents of trustworthiness of social commerce platforms: a case of rural communities using multi group SEM & MCDM methods. Electron. Commer. Res. Appl. **62**, 101322 (2023)

Al Yakin, A., Seraj, P.M.I.: Impact of metaverse technology on student engagement and academic performance: the mediating role of learning motivation. Int. J. Computations Inf. Manufact. (IJCIM) **3**(1), 10–18 (2023)

Al-Abrrow, H., Fayez, A.S., Abdullah, H., Khaw, K.W., Alnoor, A., Rexhepi, G.: Effect of open-mindedness and humble behavior on innovation: mediator role of learning. Int. J. Emerg. Markets **18**(9), 3065–3084 (2021)

Al-Adwan, A.S., Li, N., Al-Adwan, A., Abbasi, G.A., Albelbisi, N.A., Habibi, A.: Extending the technology acceptance model (TAM) to predict university students' intentions to use metaverse-based learning platforms. Educ. Inf. Technol. 1–33 (2023)

Alam, A., Mohanty, A.: Metaverse and Posthuman animated avatars for teaching-learning process: interperception in virtual universe for educational transformation. In: Panda, M. (ed.) Innovations in Intelligent Computing and Communication: First International Conference, ICIICC 2022, Bhubaneswar, Odisha, India, December 16-17, 2022, Proceedings, pp. 47–61. Springer International Publishing, Cham (2022). https://doi.org/10.1007/978-3-031-23233-6_4

Albahri, A.S., et al.: Based on the multi-assessment model: towards a new context of combining the artificial neural network and structural equation modelling: a review. Chaos Solitons Fractals **153**, 111445 (2021)

AL-Fatlawey, M.H., Brias, A.K., Atiyah, A.G.: The role of strategic behavior in achievement the organizational excellence analytical research of the manager's views of Ur state company at Thi-Qar governorate. J. Adm. Econ. **10**(37), 48–68 (2021).

Al-Hchaimi, A.A.J., Sulaiman, N.B., Mustafa, M.A.B., Mohtar, M.N.B., Hassan, S.L.B.M., Muhsen, Y.R.: A comprehensive evaluation approach for efficient countermeasure techniques against timing side-channel attack on MPSoC-based IoT using multi-criteria decision-making methods. Egypt. Inf. J. **24**(2), 351–364 (2023)

Alnoor, A., et al.: How positive and negative electronic word of mouth (eWOM) affects customers' intention to use social commerce? A dual-stage multi group-SEM and ANN analysis. Int. J. Hum. Comput. Interact. 1–30 (2022).

Alsalem, M.A., et al.: Rise of multi attribute decision-making in combating COVID-19: a systematic review of the state-of-the-art literature. Int. J. Intell. Syst. **37**(6), 3514–3624 (2022)

Asad, M.M., Malik, A.: Cybergogy paradigms for technology-infused learning in higher education 4.0: a critical analysis from global perspective. Education+Training (2023)

Atiyah, A.G.: The effect of the dimensions of strategic change on organizational performance level. PalArch's J. Archaeol. Egypt/Egyptology **17**(8), 1269–1282 (2020)

Atiyah, A.G.: Strategic network and psychological contract breach: the mediating effect of role ambiguity. Int. J. Res. Manag. Stud. (IJRMS) **13**(1), March 2023

Atiyah, A.G., Zaidan, R.A.: Barriers to using social commerce. In Artificial Neural Networks and Structural Equation Modeling: Marketing and Consumer Research Applications, pp. 115–130. Singapore: Springer Nature Singapore (2022). https://doi.org/10.1007/978-981-19-6509-8_7

Awotunde, J.B., Ayo, E.F., Ajamu, G.J., Jimoh, T.B., Ajagbe, S.A.: The influence of industry 4.0 and 5.0 for distance learning education in times of pandemic for a modern society. In: Advances in Distance Learning in Times of Pandemic, pp. 177–214. Chapman and Hall/CRC (2023)

Boletsis, C., Cedergren, J.E.: VR locomotion in the new era of virtual reality: an empirical comparison of prevalent techniques. Adv. Hum. –Comput. Interact. **2019**, 7420781 (2019)

Bozanic, D., Tešić, D., Puška, A., Štilić, A., Muhsen, Y.R.: Ranking challenges, risks and threats using Fuzzy Inference System. Decis. Mak. Appl. Manag. Eng. **6**(2), 933–947 (2023)

Cao, Y., Ng, G.W., Ye, S.S.: Design and evaluation for immersive virtual reality learning environment: a systematic literature review. Sustainability **15**(3), 1964 (2023)

Chen, X., Zou, D., Xie, H., Wang, F.L.: Metaverse in education: contributors, cooperations, and research themes. IEEE Trans. Learn. Technol. **PP**(99), 1-18 (2023)

Chew, X., Khaw, K.W., Alnoor, A., Ferasso, M., Al Halbusi, H., Muhsen, Y.R.: Circular economy of medical waste: novel intelligent medical waste management framework based on extension linear Diophantine fuzzy FDOSM and neural network approach. Environ. Sci. Pollut. Res. **30**, 1–27 (2023)

Chua, H.W., Yu, Z.: A systematic literature review of the acceptability of the use of Metaverse in education over 16 years. J. Comput. Educ. 1–51 (2023)

Creswell, J.W.: Education research planning, conducting, and evaluating quantitative and qualitative research (6thEd.). Boston: Pearson Publication (2018)

Eneizan, B., Mohammed, A.G., Alnoor, A., Alabboodi, A.S., Enaizan, O.: Customer acceptance of mobile marketing in Jordan: an extended UTAUT2 model with trust and risk factors. Int. J. Eng. Bus. Manag. **11**, 1–10 (2019). https://doi.org/10.1177/1847979019889484

Fadhil, S.S., Ismail, R., Alnoor, A.: The influence of soft skills on employability: a case study on technology industry sector in Malaysia. Interdiscip. J. Inf. Knowl. Manag. **16**, 255 (2021)

Gatea, A.A., Marina, V.: Higher education funding in Iraq in terms of the experience of particular developed countries. Int. J. Adv. Stud. **6**(1), 8–17 (2016)

Hamid, R.A., et al.: How smart is e-tourism? A systematic review of smart tourism recommendation system applying data management. Comput. Sci. Rev. **39**, 100337 (2021)

Han, J., Liu, G., Gao, Y.: Learners in the Metaverse: a systematic review on the use of roblox in learning. Educ. Sci. **13**(3), 296 (2023)

Hodges, L.C.: Teaching Undergraduate Science: A Guide to Overcoming Obstacles to Student Learning. Taylor & Francis (2023)

Kaddoura, S., Al Husseiny, F.: The rising trend of Metaverse in education: challenges, opportunities, and ethical considerations. PeerJ Comput. Sci. **9**, e1252 (2023)

Kaldarova, B., et al.: Applying game-based learning to a primary school class in computer science terminology learning. In: Frontiers in Education (Vol. 8, p. 1100275). Frontiers (2023)

Khaw, K.W., et al.: Modelling and evaluating trust in mobile commerce: a hybrid three stage Fuzzy Delphi, structural equation modeling, and neural network approach. Int. J. Hum.-Comput. Interact. **38**(16), 1529–1545 (2022)

Koohang, A., et al.: Shaping the metaverse into reality: Multidisciplinary perspectives on opportunities, challenges, and future research. Journal of Computer Information Systems (2023)

Lee, A.: Virtually Vygotsky: using technology to scaffold student learning. Technol Pedagogy, vol. 20 (2014)

Manhal, M., Al-khalidi, A., Hamad, Z.: Strategic network: managerial myopia point of view. Manag. Sci. Lett. **13**(3), 211–218 (2023)

Mitsea, E., Drigas, A., Skianis, C.: VR gaming for meta-skills training in special education: the role of metacognition, motivations, and emotional intelligence. Educ. Sci. **13**(7), 639 (2023)

Muhsen, Y.R., Husin, N.A., Zolkepli, M.B., Manshor, N., Al-Hchaimi, A.A.J.: Evaluation of the routing algorithms for NoC-Based MPSoC: a fuzzy multi-criteria decision-making approach. IEEE Access **11**, 102806–102827 (2023)

Muthmainnah, M., Khang, A., Al Yakin, A., Oteir, I., Alotaibi, A.N.: An innovative teaching model: the potential of metaverse for English learning. In: Handbook of Research on AI-Based Technologies and Applications in the Era of the Metaverse, pp. 105–126. IGI Global (2023)

Ning, H., et al.: A Survey on the Metaverse: The state-of-the-art, technologies, applications, and challenges. IEEE Internet Things J. **10**(16), 14671–14688 (2023)

Perisic, A., Perisic, I., Lazic, M., Perisic, B.: The foundation for future education, teaching, training, learning, and performing infrastructure-the open interoperability conceptual framework approach. Heliyon **9**(6), e16836 (2023)

Salloum, S., et al.: Sustainability model for the continuous intention to use metaverse technology in higher education: a case study from Oman. Sustainability **15**(6), 5257 (2023)

Yilmaz, M., O'farrell, E., Clarke, P.: Examining the training and education potential of the metaverse: results from an empirical study of next generation SAFe training. J. Softw. Evol. Process **35**(9) e2531 (2023)

Zhang, Q.: Secure preschool education using machine learning and metaverse technologies. Appl. Artif. Intell. **37**(1), 2222496 (2023)

Zhu, M., Zhang, K.: Promote collaborations in online problem-based learning in a user experience design course: Educational design research. Educ. Inf. Technol. **28**(6), 7631–7649 (2023)

Visualization Creativity through Shadowing Practices Using Virtual Reality in Designing Digital Stories for EFL Classroom

Muthmainnah[1]([✉]), Luís Cardoso[2], Ahmad Al Yakin[3], Ramsiah Tasrruddin[4], Mardhiah Mardhiah[4], and Muhammad Yusuf[5]

[1] Teacher Training and Education Faculty, Indonesia Language Department, Universitas Al Asyariah Mandar, Polewali Mandar, Indonesia
muthmainnahunasman@gmail.com
[2] Polytechnic Institute of Portalegre and Centre for Comparative Studies, University of Lisbon, Lisbon, Portugal
lmcardoso@ipportalegre.pt
[3] Teacher Training and Education Faculty, Pancasila and Civic Education Study Program, Universitas Al Asyariah Mandar, Polewali Mandar, Indonesia
ahmadalyakin76@gmail.com
[4] Dakwah dan Komunikasi, Universitas Islam Negeri Alauddin Makassar, Makassar, Sulawcsi Selatan 92113, Indonesia
mardhiah.hasan@uin-alauddin.ac.id
[5] Sekolah Tinggi Agama Islam Darud Da'wah Wal-Irsyad Makassar, Kota Makassar, Indonesia

Abstract. Findings from a quantitative study are presented here regarding the effects of VR technology on oral practices and creativity in designing digital stories of Indonesian EFL students. The population of this study was 350 undergraduate students in advanced English courses and as many as 48 students participated in an experiment learning English with a controlled Metaverse approach where they observed material on the CoSpaces Edu platform by pretending to travel by checking in at the airport, become a museum guide, solve mysteries in the castle, defeat vampires and design their own digital stories using the CoSpace Edu virtual reality (VR) application. Their perceptions were obtained from a questionnaire survey, namely the effect of VR on English acquisition, and directed observation was used to collect data processed using the statistical application SPSS version 26. The results of this study indicate the use of the Metaverse approach in language classes, which is obtained from quantitative data indicating that practice the language that experienced an increase, namely micro-skills and macro skills appeared better more actively with students' group mates with the mean score 3 agree category on the impact of VR and creativity in designing digital stories also increased with mean score 3 high creative category. Designing digital stories with the Metaverse approach, motivation, engagement, teamwork, self-confidence, and learning excitement increase as revealed by the qualitative results during learning in a hybrid environment class at university.

Keywords: Metaverse approach · Virtual Reality · Creative thinking · EFL and higher education

M. Al-Emran et al. (Eds.): IMDC-IST 2024, LNNS 895, pp. 49–60, 2023.
https://doi.org/10.1007/978-3-031-51716-7_4

1 Introduction

The term "metaverse" emerged when learning shifted to virtual space and was used to describe a hybrid online space that combines elements of augmented reality (AR) and VR (Mitra, 2023). Universities can do everything from design metaverse-based teaching materials such as VR platforms and concerts in these virtual environments to enhance 20th-century skills. Users can also make new acquaintances and forge relationships by using their online persona to interact with others. This means that they can engage in synchronous and anonymous communication and collaboration in a virtual environment under the guise of the avatar they have created. The Metaverse can be used as a productive setting for language acquisition because of its ability to encourage contact, communication, and collaboration (Aydın, 2022). Learning a language in a communicative and interactive setting is made possible by the fact that players in this virtual world must actively engage with one another and trigger student-based learning. Peterson et al. (2023) argues that foreign language students can gain a deeper understanding of the target language and culture by increasing their exposure to comprehensible input, negotiation of meaning, and output in virtual settings. Furthermore, through language exploration and creation, they can hone their problem-solving and critical thinking skills. Motivation to learn languages on their own and in groups can be encouraged through student-led discussion. As a result, Metaverse provides foreign language students with an environment that focuses on the students' own needs, allowing for independent study and group projects (Wu et al., 2023).

There are three advantages of using Metaverse to learn foreign languages (English foreign language, EFL), as stated by (Godwin-Jones, 2023) to get started, students can maintain their anonymity and engage in anonymous conversations by designing an avatar to represent themselves or as a character representing themselves. Students who are usually more withdrawn, insecure and shy may benefit from this. Second, students can practice their communication skills in a simulated real-world setting by taking on different roles in a metaverse-based virtual world. Their exchanges are lively, organic, and realistic, as opposed to static dialogue. Third, students can study online and interact with students from other countries outside the classroom walls, widening their exposure and ability to communicate with people from various cultural backgrounds. The problem with second language acquisition practice (SLA) is the need for active interaction between learners (Coumel et al., 2023). The importance of interaction in language acquisition has been known for a long time. According Courtney (2001) there is an indirect relationship between "negotiation" and acquisition because linguistic/conversational changes help in understanding input, the learning experience gained is more bound to meaning and thus helps in maximum acquisition when compared to no practice.

Much research has focused on input and interaction in SLA regarding the importance of experiential language practice. Early studies show promise, but how these interactions affect learning remains a challenge (Abbas et al., 2023; Bozanic et al., 2023). The technical difficulties in implementing interaction approaches and learning models that support classroom activities have been the subject of much research. Issues such as how to design instructional models, types of interactions and categorize student engagement and how to manage various types of variations in classroom interactions are still under investigation (Gebre et al., 2014). Seeing how interaction is a complex and dynamic activity that includes many elements, the relationship between interaction and SLA is difficult to measure. The advent of VR technology presented a way to make up for the lack of an ideal real-world setting for acquiring a second language in EFL. VR creates believable environments where students can engage with digital representations of real-world environments and people (Qiu et al., 2023). VR offers students a living environment for learning English, which has been shown to improve retention and performance in subjects (Hoang et al., 2023). In addition, VR technology has the potential to increase student interest in learning (Mubarok et al., 2023). The presence of VR platforms such as CoSpace Edu that use 360-degree spherical photos or videos to present educational materials has many advantages over 3D animation, including significant reductions in time and costs to make VR content or digital stories easier (Zhao et al., 2023). Additionally, creating content or designing digital stories does not require advanced technical skills, which suggests that many educators may be able to create their own educational materials without outside help (Muhsen et al., 2023; Al-Hchaimi et al., 2023; Chew et al., 2023). The fear of students speaking or interacting in public or among their teamwork was reduced in previous studies because they were exposed to lecture halls using VR, thereby triggering active interaction among undergraduate students. To develop students' capacities for critical thinking, creativity, and self-reflection, research has high-lighted the importance of allowing students to share their thoughts and opinions (Bi-zami et al., 2023: Atiyah, 2020). That is why it is so important to incorporate effective pedagogical practices into students' English teaching processes through technological means. In this study, Metaverse is one method that encourages students to share their creativity by analysing the VR contents and providing their new digital stories; this, in turn, improves students' creative thinking by shadowing English practices. Students participate in the Metaverse teaching approach process by submitting their own digital storytelling, assessing the work of their teams, and offering constructive criticism and suggestions. Meanwhile, they are given feedback from their peers regarding their strengths and areas for improvement. Teamwork or grouping in the hybrid learning environment has been shown to have a positive effect on students' willingness to learn, creative thinking, and reduce anxiety about English.

This study presents Metaverse teaching model, and an experiment was conducted to examine the effect of this Metaverse approach on students' English practice and their creativity in designing their own digital stories in a hybrid learning environment. Therefore, this study aims to enable learners who need to use specialized English to tackle a particular field to be motivated to learn by providing Metaverse-based technologies to unlock the most significant barriers to proficiency for highly specialized English learners. Recent advances in VR technology have presented a promising way to overcome this barrier and promise an interesting and fun learning environment that will be investigated by formulating the question, what are EFL students' perspectives of using a VR platform to design digital stories that are used to enhance their English practices and creativity in a hybrid environment class?

2 Method

2.1 Participants

This study examined the impact of VR on improving the interaction of EFL students when designing their digital stories. The population of this study is 350 students, and the sample taken by purposive sampling consisted of 48 students from computer science faculty, with 30 males (63.8%) and 17 females (36.2%). To improve their interaction in class and demonstrate their creativity in designing digital stories, eight-week courses were applied to VR in the hybrid learning environment. They were asked to observe the English materials based on the VR platform and analyse the components of the stories with their friends or their team. After six weeks (meeting 9th-14th) of observation and practicing their English, two meetings were used to design their digital stories on the VR platform. Their interaction or communication activities are recorded by observation and recording tools. A quantitative method was used to analyse the impact of VR, and the findings suggest that a metaverse learning environment can improve their interaction and creativity in designing digital storytelling more than traditional teaching modes (Figs. 1 and 2).

Fig. 1. Students digital story-telling and CoSpace Edu platform. https://edu.cospaces.io/Studio/Space/uPuodQMwqXdWtSK6

Fig. 2. Designing digital storytelling

2.2 Questionnaire

To assess the impact of VR on students' interaction skills and creativity in designing English digital stories, this study used a survey questionnaire using Google Forms which was distributed via WhatsApp social media. This questionnaire assessment uses a 5-point scale to indicate they strongly agree, 4 points strongly agree, 3 points for undecided, 2 points for disagree and 1 point for the very disagree category for 16 questions. All respondents answered the questionnaire after the English language learning treatment using VR was carried out. Every undergraduate student document their creative digital story on a YouTube account where lecturers and friends can record their language practice progress and language acquisition during this activity.

3 Data Analysis

Survey responses were collected for this investigation using SPSS 26.0. Forty-eight students answered a survey about their experience of practicing English with the virtual world in designing their digital stories. Calculation of frequencies and percentages, as well as categorization into overarching themes, is carried out using data collected from student responses. Strongly disagree is indicated by a score between 1.00 and 1.80, Strongly disagree, between 1.8 and 2.60 disagree, between 2.6 and 3.40, agree between 3.41 and 4.20, and strongly agree, between 4.21 and 5.00, agreement levels by (Pallant, 2020). Creativity categories, very high creative score 40.21–50, highly creative 30.40–40.20, moderate creative 20.61–30, low creative, 1.80–20.60 and not creative 10–10.80.

This study intends to examine the effect of the Metaverse approach on English language practices and creativity in designing digital storytelling on VR platforms. The main objective of this study was to find out students' perceptions about the Metaverse effect by analysing their perceptions using surveys and observations while using the VR platform in class. While individuals in the Metaverse group had the opportunity to hone their English practice in a synthetic environment. The impact of this VR is analysed using quantitative data from questionnaires in this VR group.

Table 1 describes the data from a student survey regarding the effect of the Metaverse Approach on their English practice during learning using the VR CoSpace Edu platform as a learning resource. The results of the data show that based on the average value obtained from the survey, which was analyzed using the SPSS version 26 statistical application, it is known that the average value is a value of 3, which indicates that all students choose the category that agrees with the Metaverse-based learning approach. This shows a strong belief in the benefits of VR in a more exhilarating hybrid learning environment, increasing enthusiasm for learning English compared to learning without the Metaverse approach, and learning is more fun with interesting VR features and games. They believe their language practice is getting better at micro (grammar and vocabulary) and macro skills (speaking, writing, listening, and reading skills) in the classroom.

Furthermore, Table 2 describes the data from the student survey regarding the effect of the Metaverse Approach on their creativity. They design digital storytelling using the VR CoSpace Edu platform as a learning resource. The results of the data show that based on the average value obtained from the survey, which was analyzed using the SPSS version 26 statistical application, it is known that the average value is a value of 3, which indicates that all students choose the category that agrees with the Metaverse-based learning

Table 1. Mean score and standard deviation on the Metaverse approach in dealing with English Practices.

Statements	N	Mean	SD
1. I didn't know about Virtual Reality before the lecturer introduced and taught us about VR applications	48	3.896	1.036
2. I like learning English with the Virtual Reality application because the design is attractive and the stories are simple and easy to understand	48	3.854	0.922
3. Using VR adds to my enthusiasm for learning English	48	3.813	0.891
4. Using VR improves my vocabulary and grammar	48	3.646	1.000
5. Using VR improves reading, writing, speaking, and listening skills	48	3.833	0.953
6. Using VR increases my confidence in learning English	48	3.792	1.071
7. I believe that by using VR, my overall motivation and interest in learning will increase	48	3.708	1.071
8. I like learning English with VR	48	3.896	0.905

Table 2. Mean score and standard deviation on Metaverse approach in dealing with Creativity in designing digital storytelling.

Items	N	Mean	Std. Deviation
1. Do you agree that VR is used in the classroom as a learning tool?	48	3.771	0.928
2. What do you think about using VR to design stories in the classroom?	48	3.604	1.233
3. What do you think about VR features in designing stories in the classroom? (character, setting)	48	3.542	1.254
4. Do you find it difficult to design stories in VR?	48	3.292	1.237
5. Do you have fun designing stories in VR?	48	3.667	1.098
6. Do you enjoy designing stories in VR that are interesting and challenging?	48	3.729	1.047
7. Do you feel uncomfortable designing stories in VR?	48	3.708	6.046
8. Did you design an English story using a VR application to add inspiration to other stories?	48	3.729	0.962

approach and can increase their creativity in designing their digital storytelling. This shows that designing stories based on a metaverse approach is interesting, challenging, and fun. They design the story, starting with determining the topic, setting, and characters, and then writing their story script in English. This shows that creativity has also been increased by the benefits of VR in a more hybrid learning environment.

4 Discussion

Universities are faced with significant barriers as they try to adapt to new digital standards in the classroom. Indonesian educators still find it difficult to assume responsibility for the successful academic advancement of 21st-century skill-based undergraduate students. Mastery of communication and creativity skills is very important to undergraduate students' careers once they have completed their education at university. Therefore, designing learning activities or learning models based on 21st-century skills is urgently needed. The results of this study have shown that students love and value this unique opportunity. This study explores the potential of VR for use in EFL classrooms. Study results show that CoSpace Edu is seen as more useful to use than images. The maximum score of the responses achieved by the students who were taught with the help of the reality system the use of VR systems assists the user in understanding abstract ideas and complex procedures. The ability to visualize material as it would in real life in a way that facilitates memory is considered a major benefit of simulation systems. However, students reported having little experience designing their digital stories due to internet network constraints or due to changes in material retention, the acquisition of new knowledge, or understanding between the two approaches. The level of simple interactivity makes students find the VR system more interactive when used as a medium and technology for learning English compared to textbooks.

The learning activities, which include small group discussions in a hybrid learning environment, trigger self-confidence to demonstrate their knowledge, stimulate creativity and joy, and involve them actively (Table 1). Moreover, the incorporation of technology into the classroom, which has become increasingly important in modern times, has facilitated the teaching and learning of the target language. Because it provides for more information than a textbook, it was time for aspiring university English instructors to implement the idea of bringing technology into the classroom. In addition, technology can assist language learning by helping students acquire and improve their language skills (Table 1) and by assisting teachers in addressing the needs of their students. As a result of the positive impact of designing learning environments with the Metaverse approach, educators now have many tools at their disposal to develop a more engaging and rewarding classroom environment for their students. In this study, researchers used the CoSpace Edu VR application for six meetings to awaken their creativity in designing digital storytelling (Table 2). In addition, the current research results confirm that gamification produces real-world experiences according to the scaffolding concept, where the Metaverse approach really helps to provide essential resources with easy access whenever and wherever they are needed, and that VR allows students to explore and interact through a variety of skills, social action, and creativity. This presents new avenues for students to express their individuality, which in turn gives them a new perspective on learning the target language with confidence (Atiyah and Zaidan, 2022).

The results show that English practice effort and creativity increase with the Meta-verse approach. Collaborative involvement with the works team, self-confidence, high reading, writing, speaking and listening skills are the effects of implementing VR in a hybrid classroom. Students participate actively in the learning process by practicing their language skills, collaborating anonymously, thinking critically and creatively (Albahri et al., 2021; Fadhil et al., 2021; Alnoor et al., 2022; Manhal et al., 2023; Atiyah, 2023). Their English skill level increases as a result. Most students have a solid grammar and vocabulary knowledge base, as well as the ability to effectively organize and manage their own learning in a virtual environment. The findings from this study indicate that the participants' knowledge and acquisition of the target language and creative settings increased as a result of their active participation in designing, theming, planning, ques-tioning, collaborating, seeking knowledge, chronology, thinking creatively, reasoning, and strategically evaluating their own learning progress (Table 2). The first investigation is intended to investigate how the Metaverse approach influences the development of language acquirer students when they learn English as a foreign language (Table 1). Researchers found that participants in the treatment group actively monitored their own learning, made their own judgments, negotiated the meaning of new material, and col-laborated electronically. To ensure more credible learning outcomes, emphasis is placed on anonymous interaction and YouTube-based performance of their language practice. This study rein-forces previous research which found that the Metaverse approach can improve English skills and students' learning awareness is positively influenced by engaging in Metaverse learning activities. The findings in this study nearly identical to some scholars who found that Metaverse, increasing (Khan et al., 2023; Alsalem, 2022), Interaction (Wang & Lai, 2023; Khaw, 2022), Creativity, (Hwang, 2023; Al-Abrrow et al., 2021); (Wu et al., 2023; Eneizan et al., 2019), English proficiency, (Kim et al., 2023), fun and happy learning, (AlSaleem et al., 2023; Hamid et al, 2021).

Although during learning the utility of VR systems is severely hampered by a lack of substantial engagement, as it causes the user's attention to wander for a longer period of time and leaves less room for learning from direct experience, it is argued that interaction is so important to users as it gives them agency. On the virtual things they do, discussing with their group mates or teams the material and play sessions that are enjoyed while interacting with VR. Students in this study can take advantage of the objects they observe in cyberspace because CoSpace Edu allows a fairly high level of object manipulation. Additionally, some students felt that the level of immersion they achieved was sufficient and that they were able to maintain their attention while immersed in the virtual envi-ronment (Table 1). The variety of content available on the CoSpace Edu platform has influenced these results by occasionally pushing students out of the virtual environment and being able to achieve the necessary level of immersion to make the experience feel truly real to participants. Some of the technical issues that occurred during the proce-dure could lead to reduced immersion, such as a slow network. They feel the Metaverse approach and virtual expeditions to imagine the physical are quite comfortable to use, such as the tutorials at airports, museums, and castles contained in the platform, which are very helpful for undergraduate students who have not yet had experience in airports, museums, and castles. There's no denying, however, that students are highly engaged with the incorporation of VR into EFL lectures and advocate its wider application in the classroom. These reactions are completely plausible, given that the main benefits of VR in education are increased interest, motivation, and creativity in the topics being taught. Students' optimistic outlook will help their education because people learn best when engaged in activities that arouse their curiosity, are actively involved, and have positive, meaningful experiences.

The potential for VR in the hybrid environment class is truly limitless. The theoretical and practical implications of VR in higher education institutes are discussed in this research, making it an invaluable resource for educators and policymakers in Indonesia. In particular, the great potential of using VR as a teaching and learning aid in English as a Foreign Language (EFL) classrooms was uncovered, which prompted more studies in VR-based education. VR-based education has the potential to radically change conventional text-based pedagogical methods by introducing new, modern opportunities for active student engagement and practical application. Students in today's heterogeneous cultures who wish to develop English language skills will greatly benefit from independently developing a strong command of a second language. Consequently, the implementation of VR can help students understand abstract concepts better, leading to more effective, realistic, exhilarating, engaging and enjoyable learning experiences. It was in line with research by (Al Yakin & Seraj, 2023; AL-Fatlawey, 2021; Gatea and Marina, 2016); Khang et al. (2023) revealed incorporating metaverse technology has been shown to increase student engagement, which in turn boosts learning outcomes. Policymakers in Indonesia's tertiary education sector should pay attention to the advantages mentioned above of VR in the classroom and the implications for Merdeka curriculum reform given the growing need for digital skills in the Industrial market. Thus, the new curriculum should (a) encourage more experimentation and hands-on, hands-on experience learning. (b) give students access to various Metaverse approaches, such as VR applications, that can help them visualize information, create information into a new concept, evaluate information, and better understand abstract and complex ideas that support 21st century skills; and (c) prepare educators and students to adapt to the Metaverse learning environment.

5 Conclusion

Digital-based learning needs can be met by integrating the Metaverse approach into a hybrid class. The results of this study show the importance of the metaverse approach that is implemented with the aim of overcoming the problems of language acquisition and creativity. This research has implications for lecturers in tertiary institutions to design learning models that maximize active practice so that language acquisition can be fulfilled and decide on game-based learning activities where students are required to practice their creativity. When choosing a VR application to use in a game-based approach to language acquisition, a teacher should consider the following factors: the purpose of the game, interesting material, clear instructions of difficulty level, high internet connectivity, available resources, and the amount of time available. It is also suggested that future research look at how English educators, based on their learning objectives, can utilize the Metaverse approach to implement blended learning. VR is demonstrating effectiveness in improving students' academic achievement; several VR-based English teaching tools are now available for use in colleges. In addition, this research offers new perspectives and a deeper understanding of how VR-based Metaverse technology needs to be utilized to enhance students' language learning abilities, particularly with regard to cultivating their creativity and increasing their participation and motivation to learn the targeted language. In conclusion, the findings show that students' optimistic views and enthusiasm for learning English efficiently in the 21st century can be carried out in a virtual classroom environment.

References

Abbas, S., et al.: Antecedents of trustworthiness of social commerce platforms: a case of rural communities using multi group SEM & MCDM methods. Electron. Commer. Res. Appl. **62**, 101322 (2023)

Al Yakin, A., Seraj, P.M.I.: Impact of metaverse technology on student engagement and academic performance: the mediating role of learning motivation. Int. J. Comput. Inf. Manufact. (IJCIM) **3**(1), 10–18 (2023)

Al-Abrrow, H., Fayez, A.S., Abdullah, H., Khaw, K.W., Alnoor, A., Rexhepi, G.: Effect of open-mindedness and humble behavior on innovation: mediator role of learning. Int. J. Emerg. Markets **18**(9), 3065–3084 (2021)

Albahri, A.S., et al.: Based on the multi-assessment model: towards a new context of combining the artificial neural network and structural equation modelling: a review. Chaos Solitons Fractals **153**, 111445 (2021)

AL-Fatlawey, M.H., Brias, A.K., Atiyah, A.G.: The role of strategic behavior in achievement the organizational excellence "analytical research of the manager's views of Ur state company at Thi-Qar governorate". J. Adm. Econ. **10**(37), 48–68 (2021)

Al-Hchaimi, A.A.J., Sulaiman, N.B., Mustafa, M.A.B., Mohtar, M.N.B., Hassan, S.L.B.M., Muhsen, Y.R.: A comprehensive evaluation approach for efficient countermeasure techniques against timing side-channel attack on MPSoC-based IoT using multi-criteria decision-making methods. Egyptian Inf. J. **24**(2), 351–364 (2023)

Alnoor, A., et al.: How positive and negative electronic word of mouth (eWOM) affects customers' intention to use social commerce? A dual-stage multi group-SEM and ANN analysis. Int. J. Hum. –Comput. Interact. 1–30 (2022).

Alsalem, M.A., et al.: Rise of multi attribute decision-making in combating COVID-19: a systematic review of the state-of-the-art literature. Int. J. Intell. Syst. **37**(6), 3514–3624 (2022)

Atiyah, A.G.: The effect of the dimensions of strategic change on organizational performance level. PalArch's J. Archaeol. Egypt/Egyptology **17**(8), 1269–1282 (2020)

Atiyah, A.G.: Strategic network and psychological contract breach: the mediating effect of role ambiguity. Int. J. Res. Manag. Stud. (IJRMS) **13**(1), March 2023

Atiyah, A.G., Zaidan, R.A.: Barriers to using social commerce. In: Artificial Neural Networks and Structural Equation Modeling: Marketing and Consumer Research Applications, pp. 115–130. Singapore: Springer Nature Singapore (2022). https://doi.org/10.1007/978-981-19-6509-8_7

Aydın, S.: The metaverse in foreign language learning: A theoretical frame-work (2022)

Bizami, N.A., Tasir, Z., Kew, S.N.: Innovative pedagogical principles and technological tools capabilities for immersive blended learning: a systematic literature review. Educ. Inf. Technol. **28**(2), 1373–1425 (2023)

Bozanic, D., Tešić, D., Puška, A., Štilić, A., Muhsen, Y.R.: Ranking challenges, risks and threats using fuzzy inference system. Decis. Mak. Appl. Manag. Eng. **6**(2), 933–947 (2023)

Chew, X., Khaw, K.W., Alnoor, A., Ferasso, M., Al Halbusi, H., Muhsen, Y.R.: Circular economy of medical waste: novel intelligent medical waste management framework based on extension linear Diophantine fuzzy FDOSM and neural network approach. Environ. Sci. Pollut. Res. **30**, 1–27 (2023)

Coumel, M., Ushioda, E., Messenger, K.: Second language learning via syntactic priming: investigating the role of modality, attention, and motivation. Lang. Learn. **73**(1), 231–265 (2023)

Courtney, M.J.: Tasks, talk and teaching: task-based language learning and the negotiation of meaning in oral interaction (2001)

Eneizan, B., Mohammed, A.G., Alnoor, A., Alabboodi, A.S., Enaizan, O.: Customer acceptance of mobile marketing in Jordan: an extended UTAUT2 model with trust and risk factors. Int. J. Eng. Bus. Manag. **11**, 1–10 (2019). https://doi.org/10.1177/1847979019889484

Fadhil, S.S., Ismail, R., Alnoor, A.: The influence of soft skills on employability: a case study on technology industry sector in Malaysia. Interdiscip. J. Inf. Knowl. Manag. **16**, 255 (2021)

Gatea, A.A., Marina, V.: Higher education funding in Iraq in terms of the experience of particular developed countries. Int. J. Adv. Stud. **6**(1), 8–17 (2016)

Gebre, E., Saroyan, A., Bracewell, R.: Students' engagement in technology rich classrooms and its relationship to professors' conceptions of effective teaching. Br. J. Edu. Technol. **45**(1), 83–96 (2014)

Godwin-Jones, R.: Emerging spaces for language learning: AI bots, ambient intelligence, and the metaverse (2023)

Hamid, R.A., et al.: How smart is e-tourism? A systematic review of smart tourism recommendation system applying data management. Comput. Sci. Rev. **39**, 100337 (2021)

Hoang, D.T.N., McAlinden, M., Johnson, N.F.: Extending a learning ecology with virtual reality mobile technology: oral proficiency outcomes and students' perceptions. Innov. Lang. Learn. Teach. **17**(3), 491–504 (2023)

Hwang, Y.: When makers meet the metaverse: effects of creating NFT metaverse exhibition in maker education. Comput. Educ. **194**, 104693 (2023)

Issa Ahmad AlSaleem, B.: The efficiency of metaverse platforms in language learning based on Jordanian young learners' perceptions. Arab World English J. (AWEJ) **14**(1), 334–348 (2023)

Khan, R.M.I., Ali, A., Kumar, T., Venugopal, A.: Assessing the efficacy of augmented reality in enhancing EFL vocabulary. Cogent Arts Humanit. **10**(1), 2223010 (2023)

Khang, A., Muthmainnah, M., Seraj, P.M.I., Al Yakin, A., Obaid, A.J.: AI-Aided Teaching Model in Education 5.0. In: Handbook of Research on AI-Based Technologies and Applications in the Era of the Metaverse, pp. 83–104. IGI Global (2023)

Khaw, K.W., et al.: Modelling and evaluating trust in mobile commerce: a hybrid three stage Fuzzy Delphi, structural equation modeling, and neural network approach. Int. J. Hum.-Comput. Interact. **38**(16), 1529–1545 (2022)

Kim, H.S., Kim, N.Y., Cha, Y.: Exploring the potential of metaverse as a future learning platform for enhancing EFL learners' English proficiency. 영어학 **23**, 220–236 (2023)

Manhal, M., Al-khalidi, A., Hamad, Z.: Strategic network: managerial myopia point of view. Manag. Sci. Lett. **13**(3), 211–218 (2023)

Mitra, S.: Metaverse: a potential virtual-physical ecosystem for innovative blended education and training. J. Metaverse **3**(1), 66–72 (2023)

Mubarok, H., Lin, C.J., Hwang, G.J.: A virtual reality-based collaborative argument mapping approach in the EFL classroom. Interact. Learn. Environ. 1–19 (2023)

Muhsen, Y.R., Husin, N.A., Zolkepli, M.B., Manshor, N., Al-Hchaimi, A.A.J.: Evaluation of the routing algorithms for NoC-based MPSoC: a fuzzy multi-criteria decision-making approach. IEEE Access **11**, 89–106 (2023)

Pallant, J.: SPSS survival manual: A step by step guide to data analysis using IBM SPSS. McGraw-hill education (UK) (2020)

Peterson, M., Wang, Q., Mirzaei, M.S.: The use of network-based virtual worlds in second language education: A research review. Research Anthology on Virtual Environments and Building the Metaverse, pp. 218–236 (2023)

Qiu, X.B., Shan, C., Yao, J., Fu, Q.K.: The effects of virtual reality on EFL learning: a meta-analysis. Educ. Inf. Technol. 1–27 (2023)

Wang, H., Lai, P.C.: Classroom interaction and second language acquisition in the metaverse world. In: Strategies and Opportunities for Technology in the Metaverse World, pp. 186–195. IGI Global (2023)

Wu, J.G., Zhang, D., Lee, S.M.: Into the brave new metaverse: envisaging future language teaching and learning. IEEE Trans. Learn. Technol. 1–11 (2023)

Wu, W.C.V., Manabe, K., Marek, M.W., Shu, Y.: Enhancing 21st-century competencies via virtual reality digital content creation. J. Res. Technol. Educ. **55**(3), 388–410 (2023)

Zhao, W., Su, L., Dou, F.: Designing virtual reality based 3D modeling and interaction technologies for museums. Heliyon **9**(6), e16486 (2023)

Metaverse Applications and Its Use in Education

Abdullah A. Nahi[1(✉)], Arkan A. Ghaib[1], and Ahmed Abd Aoun Abd Ali[2]

[1] Department of Information Technologies, Management Technical College, Southern Technical University, Basrah, Iraq
eng.abdullahnahi@gmail.com, arkan.ghaib@stu.edu.iq
[2] Department of General Education, Cihan University, Erbil, Iraq
Ahmed.Aoun@cihanuniversity.edu.iq

Abstract. The Metaverse, an amalgamation of virtual reality (VR), augmented reality (AR), and digital technologies has captured substantial attention for its potential to reshape human interaction, collaboration, and experiences. This comprehensive review examines Metaverse and its applications across multiple domains. The paper begins by elucidating the foundational technologies of the Metaverse, such as VR, AR, and spatial computing. It traces the evolution of the concept and its significance in bridging physical and digital realms. The review delves into diverse applications of the Metaverse. It explores how the Metaverse facilitates novel forms of social engagement, immersive Education, and remote collaboration. It also assesses its impact on entertainment, commerce, healthcare, and personal expression. The review emphasizes the potential of the Metaverse to redefine workspaces, enable unique gaming experiences, and revolutionize economic models. Ethical concerns, technical challenges, and regulatory frameworks are discussed, emphasizing the need for privacy, data security, and ethical guidelines. In conclusion, the paper underscores the transformative potential of the Metaverse across industries while highlighting the necessity for ethical considerations. The ongoing evolution of the Metaverse demands continuous research and collaboration to ensure its responsible integration into society.

Keywords: Metaverse · virtual reality · augmented reality · spatial computing · social interaction · education

1 Introduction

The digital realm is undergoing a profound transformation, transcending the boundaries of traditional virtual spaces and ushering in an era of interconnected, immersive experiences. At the forefront of this technological revolution lies the concept of the Metaverse, a term coined in science fiction but increasingly becoming a tangible reality (Contreras et al., 2022; Eneizan et al., 2019). The Metaverse represents a convergence of virtual and physical worlds, where individuals can seamlessly navigate, interact, and create within a complex digital ecosystem. The genesis of the Metaverse can be traced back to the early days of the Internet. Still, it has recently gained unprecedented momentum, driven by advancements in virtual, augmented reality, blockchain technology, and artificial intelligence (González Vallejo, 2023; Hamid et al., 2021). Today, the Metaverse extends far beyond gaming and entertainment, encompassing various applications across various sectors, including commerce, healthcare, and education.

This review paper embarks on a comprehensive exploration of metaverse applications and their burgeoning role in Education. It delves into the multifaceted dimensions of the Metaverse, discussing its evolution, defining characteristics, and the technological underpinnings that power its growth. The paper will elucidate how the Metaverse has evolved from a realm predominantly occupied by gamers into a dynamic, interconnected space with implications far beyond entertainment (Al-Abrrow et al., 2021). The pivotal focus of this paper lies in unraveling the Metaverse's potential to revolutionize education. Traditional educational paradigms have faced significant disruptions in recent years, exacerbated by the global shift towards remote and online learning during the COVID-19 pandemic. The Metaverse emerges as a transformative force capable of reimagining the entire educational landscape (Zhang et al., 2022; Khaw et al., 2022). By seamlessly blending the physical and digital realms, it offers novel opportunities for engagement, collaboration, and personalized learning experiences that have the potential to transcend the limitations of traditional classrooms.

This exploration is timely as educators, policymakers, and technology developers increasingly grapple with how to harness the Metaverse's capabilities to enhance educational outcomes while addressing the associated challenges and ethical considerations (Muhsen et al., 2023; Al-Hchaimi et al., 2023; Chew et al., 2023). As we embark on this journey through the Metaverse's myriad applications and its integration into education, we aim to shed light on the promises, the perils, and the possibilities that lie ahead, ultimately paving the way for informed discourse and strategic decision-making in the realm of digital learning (Ahuja et al., 2023; Alsalem et al., 2022). In the following sections, we will traverse the metaverse landscape, examining its diverse applications across industries, before zooming in on the unique ways it is poised to reshape Education. Through an in-depth analysis of current trends, case studies, and critical discussions, we hope to provide valuable insights into the Metaverse's role in shaping the future of learning (Albahri et al., 2021).

2 Literature Review

The Metaverse, a digital space where virtual and physical worlds intersect, has garnered significant attention recently (A. Nahi et al., 2022; Fadhil et al., 2021). While initially a concept from science fiction, it has evolved into a tangible technological phenomenon. This literature review explores the growing body of research and discourse surrounding metaverse applications and their transformative potential in Education (Jameel et al., 2022; Alnoor et al., 2022). Research on metaverse applications in education has been a burgeoning field of interest. The Metaverse, often described as a collective virtual shared space, is increasingly recognized for its potential to revolutionize education. Recent studies have emphasized the immersive and interactive nature of metaverse environments, which can facilitate experiential learning. For instance, a study by (Johnson et al. 2021; AL-Fatlawey et al., 2021) investigated the integration of metaverse platforms into higher education, highlighting its effectiveness in providing students with opportunities for hands-on experiences in simulated environments. Students can engage with complex concepts in fields like engineering and medicine in a safe and controlled setting through virtual reality and augmented reality technologies (Gatea and Marina, 2016). The study also emphasized the potential for global collaboration and knowledge sharing, as learners from diverse locations can collaborate in the Metaverse to work on projects and share insights (Abbas et al., 2023; Bozanic et al., 2023).

Furthermore, another recent study by (Chen and Zhang 2022; Manhal et al., 2023) delved into the role of the Metaverse in enhancing inclusivity in education. The authors argued that metaverse platforms offer a level playing field for students with varying physical abilities and learning styles. The Metaverse can cater to individual preferences and needs through customizable avatars and adaptable interfaces, potentially reducing learning barriers (Atiyah and Zaidan, 2022). This study also highlighted the potential for accessibility features such as real-time translation and transcription services, making educational content more understandable for non-native speakers and individuals with hearing impairments. These recent studies underscore the transformative potential of metaverse applications in education, offering new avenues for immersive, inclusive, and collaborative learning experiences. These virtual shared spaces offer a dynamic and immersive environment for learning, presenting opportunities for experiential and interactive educational experiences. Studies have highlighted the integration of metaverse platforms in higher education, showcasing their effectiveness in providing students with hands-on experiences in simulated environments (Atiyah, 2023). Through the utilization of virtual reality and augmented reality technologies, students can engage with complex concepts in fields like engineering and medicine in a controlled and secure setting. Additionally, research has underscored the Metaverse's capacity for inclusivity and accessibility in education. By customizing avatars and interfaces, the Metaverse has shown promise in catering to diverse learning styles and abilities, potentially reducing barriers to learning (Atiyah, 2020). These findings collectively indicate that metaverse applications hold significant promise in revolutionizing education, offering a platform for immersive, inclusive, and collaborative learning experiences.

2.1 The Evolution of the Metaverse Concept

The notion of the Metaverse traces its roots back to science fiction, notably Neal Stephenson's 1992 novel "Snow Crash" and later works such as "Ready Player One" by Ernest Cline (Hashim et al., 2022). It represents a digital space where individuals can interact, collaborate, and create. Initially, the Metaverse was synonymous with immersive online gaming environments. However, it has transcended these origins, encompassing a broader spectrum of experiences and applications. The notion of the Metaverse carries a rich history, originating from the imaginative realms of science fiction. Neal Stephenson's 1992 novel "Snow Crash" first introduced the concept, portraying a virtual universe where individuals could converge, engage, and even conduct commerce. This vision of a digital space paved the way for subsequent works like Ernest Cline's "Ready Player One," which further delved into the immersive potential of such environments. As these narratives gained popularity, they sparked a collective imagination about the possibilities of a shared virtual realm.

Initially, the Metaverse was predominantly associated with immersive online gaming environments, where players could assume avatars, navigate fantastical landscapes and engage in quests or challenges. However, its evolution has been nothing short of remarkable. Today's Metaverse encompasses many experiences and applications that extend far beyond gaming. It is a multifaceted digital ecosystem where individuals can play, work, learn, socialize, and conduct business. The Metaverse has become a versatile platform for various human activities, from virtual meetings and conferences to educational simulations and art exhibitions. This shift signifies a significant paradigm change with implications for various domains, including education. The Metaverse's capacity for interaction, collaboration, and creation offers unprecedented opportunities for innovative pedagogical approaches. It holds the potential to revolutionize how knowledge is shared, assimilated, and applied, presenting educators with a dynamic and immersive toolset to engage learners in once inconceivable ways. As research in this field advances, it will be crucial to explore how educational practices can best leverage the Metaverse to enhance learning outcomes and foster a more inclusive and accessible educational experience for students of all backgrounds and abilities.

2.2 Technological Foundations of the Metaverse

The Metaverse's rapid development owes much to advances in technology. Key enablers include virtual reality, augmented reality, mixed reality, blockchain, and artificial intelligence. VR and AR headsets and haptic devices have made it possible to create immersive virtual environments. Blockchain technology provides security, ownership, and interoperability in the Metaverse (Hwang & Chien, 2022). AI enhances user experiences by enabling intelligent interactions within these digital realms. The Metaverse is no longer confined to the gaming sector. It has applications in various industries, including commerce, entertainment, healthcare, and social interactions. Businesses leverage metaverse technologies for virtual commerce and customer engagement (Zhang et al., 2022). Healthcare practitioners use it for simulations and telemedicine. Social platforms are evolving into metaverse spaces where people gather, socialize, and collaborate.

2.3 The Metaverse in Education: A Paradigm Shift

One of the most promising areas of metaverse application is in education. Traditional educational models have faced challenges adapting to the digital age's demands. The Metaverse offers unique solutions to these challenges. It allows for immersive and interactive learning experiences, fostering student engagement and collaboration. This technology can break geographical barriers, enabling remote and global learning communities (Lin et al., 2022). In recent years, researchers and educators have explored the integration of metaverse technologies into educational settings. Virtual classrooms, interactive simulations, and immersive historical recreations are examples. These environments allow students to learn through experiential, hands-on activities, enhancing retention and understanding.

2.4 Challenges and Considerations

While the potential benefits of metaverse integration in Education are substantial, challenges and concerns persist. Privacy issues, ethical dilemmas, accessibility barriers, and the digital divide are among the critical considerations. Ensuring equity in metaverse-enhanced Education remains a pressing concern (Kaddoura & Al Husseiny, 2023). The intersection of the Metaverse and Education is still in its nascent stages. There is a need for further research to understand the full scope of possibilities and challenges. This includes investigating practical pedagogical approaches, evaluating the impact on learning outcomes, and addressing ethical and equity concerns. The Metaverse holds great promise, but its successful implementation in education requires careful planning and research-driven insights (Chen, 2022).

3 Metaverse Applications

The Metaverse, born from the convergence of cutting-edge technologies and visionary imagination, has evolved into a versatile ecosystem with diverse applications across numerous industries. This section delves into the domains where the Metaverse is making significant inroads, showcasing its potential beyond traditional virtual environments (Kye et al., 2021).

3.1 Gaming and Entertainment

Historically, the metaverse concept emerged from the gaming industry and remains a vibrant hub for its development. Virtual worlds and massively multiplayer online games (MMOs) have evolved into intricate metaverse realms where players assume digital personas, engage in collaborative missions and even trade virtual assets using blockchain technology. These metaverse gaming experiences offer immersive, social, and dynamic environments, blurring the lines between reality and fiction (Abbate et al., 2022).

3.2 Social Interactions

Beyond gaming, the Metaverse is redefining social interactions and online communities. Metaverse platforms enable individuals to meet, converse, and collaborate within virtual spaces. Social VR environments allow face-to-face interactions, virtual conferences, and concerts, enhancing the sense of presence and connection, especially in remote or isolated settings (Zhang, 2022).

3.3 Virtual Commerce

Metaverse is reshaping the commerce landscape, offering a new frontier for businesses to engage with customers. Virtual stores and showrooms allow consumers to explore and interact with products before purchasing. NFT (Non-Fungible Token) marketplaces within the Metaverse enable the ownership and trading of digital assets, including artwork, collectibles, and virtual real estate (Kavut, 2023).

3.4 Healthcare and Therapy

In the healthcare sector, Metaverse has found applications in medical training, telemedicine, and therapeutic interventions. Medical students can practice surgical procedures in realistic simulations, while patients can receive therapy in immersive virtual environments. The Metaverse's potential to simulate real-world scenarios enhances training and therapeutic outcomes (González Vallejo, 2023).

3.5 Education and Training

Education stands out as one of the most promising domains for metaverse applications. Virtual classrooms, interactive simulations, and field trips create engaging and experiential learning environments. Students can explore historical events, conduct science experiments, or participate in collaborative projects regardless of physical location, fostering global learning communities (Araujo Inastrilla, 2023).

3.6 Business and Collaboration

Metaverse technologies are transforming the way businesses operate. Virtual offices, co-working spaces, and collaborative environments allow employees to work together seamlessly, irrespective of geographical constraints. These digital spaces facilitate meetings, brainstorming sessions, and project management, enhancing productivity and global collaboration (Araujo Inastrilla, 2023).

3.7 Art and Creativity

Artists and creators embrace the Metaverse to showcase their work and engage with audiences. Virtual art galleries and immersive experiences enable artists to reach global audiences. Additionally, metaverse tools empower creators to build interactive, multimedia-rich content, pushing the boundaries of digital artistry (Forte & Kurillo, 2010).

3.8 Research and Development

In research and development, the Metaverse offers simulations, data visualization, and collaborative environments for scientists, engineers, and innovators. The Metaverse accelerates progress in various fields, from simulating complex experiments to conducting virtual conferences. These diverse metaverse applications underscore its transformative potential, offering new ways to interact, work, learn, and create in an interconnected digital realm. As technologies evolve, Metaverse's footprint is expected to expand further, shaping how we engage with the digital world across industries and sectors (Forte & Kurillo, 2010).

4   Metaverse in Education

The Metaverse, a dynamic digital realm defined by its immersive and interconnected nature, has emerged as a promising frontier in education. It offers innovative solutions to traditional educational models' challenges, presenting opportunities to transform how knowledge is imparted, acquired, and experienced. In this section, we explore the multifaceted role of the Metaverse in education and its potential to reshape the learning landscape.

4.1   Immersive Learning Environments

One of the key strengths of the Metaverse in education lies in its ability to create immersive and experiential learning environments. Virtual and augmented reality technologies allow students to enter. These immersive experiences enhance retention, understanding, and engagement, making learning more dynamic and memorable (Bisset Delgado, 2022).

4.2   Global and Remote Learning

The Metaverse transcends physical boundaries, making it a powerful global and remote learning tool. Students worldwide can unite in virtual classrooms, breaking down geographical barriers. This fosters diverse perspectives and cultural exchanges, enriching the educational experience. Furthermore, learners in remote or underserved areas access quality education through the Metaverse (Germen, 2010).

4.3   Collaborative Learning

Collaboration is a fundamental aspect of education, and the Metaverse facilitates collaborative learning in unprecedented ways. Virtual group projects, teamwork in shared virtual spaces, and real-time assignment collaboration enable students to develop essential interpersonal and problem-solving skills. The Metaverse promotes active engagement and peer learning, fostering a sense of community among learners (Bisset Delgado, 2022).

4.4   Personalized Learning

Metaverse technologies allow for personalized learning experiences. AI-driven adaptive learning platforms can tailor content and pacing to individual student needs and preferences. Learners can explore topics at their own pace, receive real-time feedback, and access resources customized to their learning styles, optimizing the learning journey (Zhang, 2022).

4.5   Professional Development and Training

In addition to K-12 and Higher Education, the Metaverse plays a crucial role in professional development and training. Businesses and organizations use virtual simulations and environments to train employees in diverse fields, from healthcare to aviation. These realistic training scenarios enhance skills and readiness, reducing the need for physical equipment or facilities (WİSNU BUANA, 2023).

4.6   Accessibility and Inclusivity

The Metaverse offers opportunities to enhance accessibility and inclusivity in education. Text-to-speech, speech-to-text, and customizable avatars cater to diverse learning needs. Moreover, the Metaverse can provide accommodations for students with disabilities, ensuring equitable access to Education (Zhang, 2022). While the Metaverse holds excellent promise in education, it also presents challenges and considerations. Privacy concerns, digital equity, and ethical considerations, such as data security and virtual harassment, require careful attention. Educators and policymakers must address these issues to ensure safe and equitable metaverse-enhanced learning environments (Wisnu Buana, 2023).

5   Challenges and Concerns in Education

Indeed, integrating the Metaverse into education presents numerous exciting possibilities, but it also comes with challenges and concerns that need careful consideration. In this section, we explore some key challenges and concerns associated with using the Metaverse in education, as shown in Table 1.

Table 1. Challenges and Concerns in Education.

Subject	Challenge	Concern	Resources
1. Digital Equity and Access	Not all students have equal access to the hardware and internet bandwidth required for a seamless metaverse experience	Ensuring equitable access to metaverse-based Education is essential to prevent excluding students from marginalized communities who lack the necessary technology and connectivity	(Sá & Serpa, 2023)
2. Privacy and Data Security	The Metaverse collects vast amounts of user data to create immersive experiences. This data can include personal information and behavioral patterns, raising concerns about privacy and data security, particularly for children and young learners	Robust data protection measures, transparent consent processes, and strict adherence to privacy regulations are essential to protect students' sensitive information and maintain trust in metaverse platforms	(Lau, 2015)
3. Pedagogical Integration	Integrating the Metaverse into educational practices effectively can be challenging. Educators may struggle to fully adapt their teaching methods and curricula to exploit the Metaverse's capabilities	Concern: Professional development and training for teachers are essential to help them effectively utilize metaverse technologies in pedagogically sound ways, ensuring that the Metaverse enhances, rather than distracts from, the learning experience	(Carrisi, 2020)

(continued)

Table 1. (*continued*)

Subject	Challenge	Concern	Resources
4. Content Quality and Curation	The Metaverse is a vast space where user-generated content coexists with professionally developed educational materials. Ensuring educational content's quality, accuracy, and appropriateness can be challenging	Establishing guidelines and standards for educational content within the Metaverse is crucial to ensure students access high-quality, reliable materials	(Ghantous & Fakhri, 2022)
5. Ethical Considerations	The Metaverse can blur the lines between the real and virtual worlds, raising ethical questions about identity, behavior, and interaction within digital spaces	Addressing ethical concerns, such as identity theft, impersonation, or the impact of immersive experiences on mental health, is essential to ensure a safe and responsible metaverse environment for students	(Zahedi, et al., 2023)
6. Cost and Sustainability	Building and maintaining metaverse platforms and experiences can be costly. Schools and educational institutions may struggle to allocate resources for metaverse-based Education	Achieving cost-effectiveness and sustainability in metaverse education initiatives is critical to ensure they remain accessible and viable in the long term	(Garon, 2023)
7. Teacher-Student Relationship	The Metaverse can change the dynamics of the teacher-student relationship, potentially reducing face-to-face interactions and personal connections	Balancing the benefits of technology with the importance of interpersonal relationships in Education is crucial. Maintaining a sense of connection and mentorship in the Metaverse is a concern that requires careful attention	(Ryan, 2023)

As a result, while the Metaverse offers exciting opportunities to enhance education, addressing these challenges and concerns is essential for its successful integration into educational settings. A thoughtful and collaborative approach involving educators, policymakers, technology developers, and parents is needed to harness the Metaverse's potential while mitigating its associated risks in education. Indeed, the Metaverse offers many benefits and opportunities in education, promising to transform the way students learn and educators teach (Talan & Kalinkara, 2022). This section explores vital advantages and opportunities the Metaverse brings to education, as shown in Fig. 1.

The Metaverse offers many benefits for education, revolutionizing how students learn, and educators teach. First and foremost, it provides immersive learning experiences through virtual and augmented reality that engage students in ways traditional classrooms cannot. It enables exploration of historical events and complex concepts as if they were physically present, thus enhancing comprehension and retention. Additionally, the Metaverse fosters global collaboration and connectivity, breaking geographical boundaries and facilitating cultural exchange among students from diverse backgrounds.

Personalized learning is another advantage, as metaverse technologies tailored to individual needs, powered by AI, optimize the learning process by assessing strengths and weaknesses, adjusting instruction, and providing targeted feedback. Furthermore, the Metaverse addresses accessibility challenges, making learning more inclusive through features like text-to-speech, speech-to-text, and customizable avatars for students with disabilities (Akçayır & Akçayır 2017; Özdemir et al., 2022; George Reyes, 2020).

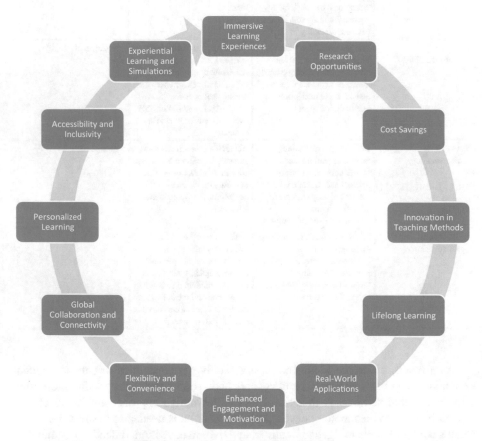

Fig. 1. Benefits of Metaverse in Education.

Experiential learning and realistic simulations in virtual laboratories enhance critical thinking, while flexibility and convenience enable access to educational content from anywhere, accommodating diverse learning styles. Gamified elements in the Metaverse enhance engagement and motivation, making learning enjoyable. It also prepares students for real-world applications of knowledge and skills, promoting practical insights relevant to future careers. Lifelong learning is supported, enabling continuous skill development, which is crucial in an era requiring adaptability for career success. Innovative teaching methods are encouraged, adapting to evolving learner needs and job market demands. Cost savings are a long-term benefit, with reduced physical infrastructure needs. Finally, the Metaverse offers rich research opportunities, enabling the study of its impact on learning outcomes and pedagogical approaches, advancing educational theory and practice. In summary, the Metaverse opens up opportunities to enhance and innovate education. Its immersive, collaborative, and personalized features can improve learning outcomes, promote inclusivity, and prepare students for the challenges and opportunities of the future. By harnessing the potential of the Metaverse, education can become more engaging, accessible, and effective.

6 Methodology

Conducting a systematic literature review on "Metaverse in education and its applications" involves a structured approach to gathering, evaluating, and synthesizing relevant academic literature; we followed these steps.

6.1 Research Question and Objectives

As shown in the following figure, we defined the research question(s) and objectives to guide the review (Fig. 2).

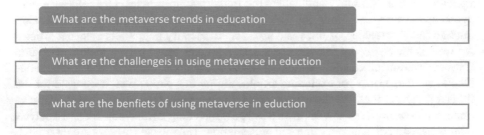

Fig. 2. Research Questions.

After we highlighted the research questions, we set the search strategy through similar keywords that could be found in the target papers that we went through from different academic databases, search engines, and specialized research repositories. Consider including published and unpublished sources, as seen in the following figures.

We also focused on the latest resources since the topic started to be developing the last few years, and for that, we conducted the Boolean operators and filters. Combine search terms using Boolean operators (AND, OR, NOT) to narrow or broaden the search. Utilize filters such as publication date (e.g., last 5–10 years) to focus on recent research. Consider language filters if applicable.

6.2   Inclusion and Exclusion Criteria

Defining explicit inclusion and exclusion criteria to determine which articles to include in the review.

Inclusion criteria include:

Articles published in peer-reviewed journals, conference proceedings, and academic books.

Focus on using the Metaverse, VR, AR, or similar immersive technologies in educational contexts.

Articles discussing applications, benefits, challenges, and outcomes of metaverse use in Education.

Research studies, case studies, reviews, and empirical studies.

For the exclusion criteria, we can see the in the following steps

1) Non-academic or non-peer-reviewed sources.
2) Irrelevant topics or studies not related to Education.
3) Articles lacking empirical data or scholarly content.
4) Studies published before a specific date (before 2010).

6.3   Data Collection and Screening

Execute the search strategy and collect the initial pool of articles. Use reference management software (e.g., EndNote, Zotero) to organize and manage the literature. Conduct an initial screening of titles and abstracts to identify potentially relevant articles. As well as reviewing the full texts of selected articles based on the inclusion criteria. Extract relevant information, including research methods, findings, and implications. Document the reasons for excluding articles during this phase, which we used for that dtSearch Indexer tool. We focused on critical notes while doing that. A systematic literature review involves several critical stages, including data synthesis and analysis, quality assessment, data reporting and presentation. The selected articles are thoroughly reviewed during the data synthesis and analysis phase, and their findings are synthesized to identify key themes, trends, and research questions. This process involves organizing the content to uncover patterns, relationships, and gaps in the literature, providing a comprehensive understanding of the research landscape. Quality assessment is a crucial step to ensure the reliability and validity of the selected studies. Factors such as research design, methodology, sample size, and data analysis techniques are carefully evaluated to gauge the quality and rigor of the research. Assessing the findings' validity and reliability helps determine the trustworthiness of the literature under review.

7   Discussion

The Metaverse, a term once confined to science fiction and speculative technology, is rapidly becoming a tangible reality. It represents a paradigm shift in how we conceive of digital spaces and interactions, opening up various opportunities and challenges across various domains. One of the most promising frontiers for the Metaverse is its role in education. In this comprehensive discussion, we will delve into the Metaverse's potential to reshape the education landscape, offering immersive learning experiences, global collaboration, personalized learning pathways, accessibility, and inclusivity as the following points.

1. Immersive Learning Experiences: Bridging the Reality Gap

One of the Metaverse's most transformative aspects is its capacity to create immersive learning experiences. Virtual reality and augmented reality technologies have reached a level of sophistication that enables students to step beyond the confines of traditional classroom settings. Through VR headsets and AR devices, learners can embark on virtual journeys to historical events, dive into scientific simulations, or even traverse the surface of Mars. The Metaverse can transform passive learning into active, experiential education. This shift has profound implications for enhancing comprehension, retention, and engagement. Imagine a history class where students find themselves standing on the battlefields of the American Civil War, witnessing the struggles and sacrifices of soldiers firsthand. Or a biology lesson where students can explore the inner workings of a human cell in 3D, interacting with its components to understand complex biological processes. These immersive experiences make learning more engaging and cater to diverse learning styles, fostering a deeper understanding of the subject matter.

2. Global Collaboration and Connectivity: Expanding Educational Horizons

The Metaverse's capacity to transcend geographical boundaries opens new vistas for global collaboration and connectivity in education. Traditional classroom walls are replaced by virtual spaces where students and educators worldwide can gather, share experiences, and collaborate seamlessly. This interconnectedness fosters a global perspective, promotes cultural exchange, and broadens students' horizons. In a metaverse-enabled educational landscape, students can participate in joint projects with peers from diverse cultural backgrounds, gaining insights into different worldviews and approaches to problem-solving. This enriches the educational experience and prepares students for the globalized, interconnected world they will encounter in their future careers.

3. Personalized Learning: Nurturing Individual Growth

The Metaverse is not a one-size-fits-all educational solution; it offers a personalized learning ecosystem tailored to individual student needs and preferences. Artificial intelligence plays a pivotal role in this customization. AI-powered adaptive learning platforms can assess each student's strengths and weaknesses, adjusting the content and pace of instruction to match their learning trajectory. Real-time feedback loops enable students to track their progress and make informed decisions about their learning pathways. Imagine a scenario where a student studying mathematics receives tailored lessons and exercises that align with their current level of proficiency. As they master concepts, the difficulty level automatically adjusts to keep them challenged but not overwhelmed. Similarly, a student with dyslexia can utilize metaverse tools that offer text-to-speech features, making learning materials more accessible. Such personalized learning experiences have the potential to address diverse learning styles and abilities, fostering greater inclusivity in education.

4. Accessibility and Inclusivity: Overcoming Educational Barriers

The Metaverse has the potential to break down traditional barriers to education and promote inclusivity. For students with disabilities, metaverse technologies offer many tools and features that can level the playing field. Screen readers, voice commands, and customizable avatars are just some accessibility features that can empower learners with disabilities. Consider a student with a visual impairment who can access educational materials through a screen reader in a metaverse-based classroom. This technology can convert text to speech, providing a seamless learning experience. Alternatively, students with limited mobility can navigate virtual environments and engage in activities using intuitive gesture-based controls, ensuring active participation in learning activities. Inclusivity in education is not solely limited to accommodating disabilities. It extends to learners from diverse cultural and socio-economic backgrounds. The Metaverse's capacity to foster global collaboration and tailor learning experiences to individual needs contributes to a more inclusive educational landscape.

5. The Metaverse: Catalyst for Lifelong Learning

Education is no longer confined to the early years of life. In an age of rapid technological advancement and changing job landscapes, the Metaverse empowers individuals to engage in lifelong learning. Lifelong learning is not a mere catchphrase; it's a necessity for thriving in a dynamic world. The Metaverse's dynamic and evolving nature aligns perfectly with the concept of lifelong learning. Individuals can continuously acquire new knowledge and skills in response to emerging technologies and shifting industry demands. Virtual workshops, interactive tutorials, and collaborative projects provide opportunities for ongoing personal and professional development. Consider a mid-career professional looking to pivot into a new field. They can immerse themselves in metaverse-based training programs, gaining hands-on experience in their chosen field and networking with experts and peers. This continuous learning journey can be pursued at their own pace, aligning with their career aspirations and evolving skill sets.

6. Innovation in Teaching Methods: Educators as Creators

The Metaverse is a tool for students and a canvas for educators to explore innovative teaching methods. Educators can craft immersive, interactive lessons that capture students' imaginations and foster critical thinking. These experiences range from historical reenactments and virtual field trips to complex scientific simulations and collaborative projects. Imagine a biology teacher using the Metaverse to take their students on a virtual dive into a coral reef ecosystem, where they can observe marine life up close and even participate in ecological research. Alternatively, a language teacher can create virtual environments where students engage in real-world conversations with native speakers, honing their language skills through immersion. The Metaverse empowers educators to become creators, designing educational experiences that align with their pedagogical goals and the needs of their students. This shift from passive consumption to active creation is at the heart of educational innovation in the Metaverse.

7. Cost Savings and Sustainable Education

While initial investments may be required to establish metaverse-based educational platforms, they can lead to cost savings in the long run. Physical infrastructure, such as school buildings, maintenance, and transportation, can be reduced as more learning occurs in virtual environments. Additionally, the Metaverse enables the sharing educational resources and expertise across institutions, optimizing resource allocation. Moreover, sustainable education is increasingly vital in an era of environmental concerns. By reducing the need for physical facilities and travel, metaverse-based education contributes to a more environmentally friendly and sustainable educational ecosystem.

8. Research Opportunities: Advancing Educational Theory and Practice

Integrating the Metaverse into education opens up a rich research and experimentation landscape. Researchers can investigate the impact of metaverse technologies on learning outcomes, pedagogical approaches, and student engagement. They can study the effectiveness of immersive learning experiences, the benefits of global collaboration, and the nuances of personalized learning in virtual environments. Such research not only furthers our understanding of how the Metaverse can enhance education but also informs best practices for educators and policymakers. It contributes to the evolution of educational theory and practice, ensuring that metaverse-based education is grounded in evidence and data-driven insights.

9. Ethical Considerations: Navigating the Metaverse Responsibly

As with any technological advancement, the Metaverse brings ethical considerations to the forefront. The seamless blend of digital and physical realities within the Metaverse raises questions about identity, behavior, and interaction. Students and educators must navigate these spaces responsibly and ethically. Identity theft, impersonation, and the potential for harassment in virtual environments demand attention. Safeguards and guidelines must be in place to protect the well-being and dignity of all participants in metaverse-based education. Educators and students should receive guidance on digital citizenship, online etiquette, and responsible behavior in virtual spaces.

10. Balancing Technology with Interpersonal Connections

While the Metaverse offers transformative technological capabilities, balancing technology and interpersonal connections in education is crucial. The teacher-student relationship is a cornerstone of effective education, and the Metaverse should complement, not replace, the importance of personal interactions. Educators must maintain a sense of mentorship, guidance, and emotional support, even within virtual learning environments. Building rapport, fostering communication, and creating a sense of belonging are essential to ensuring that the Metaverse enhances rather than detracts from the educational experience.

8 Future Work

The Metaverse's integration into education is still evolving, with ongoing research and development. Future directions include refining pedagogical approaches, measuring the impact on learning outcomes, and developing best practices for metaverse-based education. The journey toward fully realizing the Metaverse's potential in education is an ongoing endeavor driven by innovation and collaboration. Future research in metaverse applications in education holds immense promise for reshaping the landscape of learning environments. One crucial avenue of exploration lies in optimizing the pedagogical strategies within metaverse platforms. Understanding how different instructional methods and assessment techniques can be seamlessly integrated into these immersive environments will be imperative. Additionally, delving into the nuanced dynamics of student engagement and motivation within the Metaverse is a vital area of inquiry. Exploring gamification principles, collaborative learning experiences, and personalized educational pathways within these virtual realms will provide invaluable insights into how to foster a more interactive and participatory learning process. Furthermore, future studies should address the development of robust frameworks for evaluating the effectiveness of metaverse-based education. This entails establishing clear metrics for learning outcomes, cognitive engagement, and skill acquisition within virtual environments. Additionally, research should focus on the long-term impacts of metaverse-based education on students' academic performance and retention rates. Understanding the sustained benefits and potential challenges of incorporating the Metaverse into educational settings will ensure its seamless integration into mainstream educational practices. Ultimately, future research endeavors in this domain are poised to refine the efficacy of metaverse applications in education and pave the way for a more innovative and inclusive educational paradigm. In summary, the Metaverse represents a transformative force in education, offering new avenues for immersive, global, and personalized learning experiences. As educators, learners, and technology developers continue to explore its capabilities, the Metaverse has the potential to revolutionize the way we teach and learn in the digital age.

9 Conclusion

The Metaverse represents a new chapter in the history of education, offering a compelling vision of the future of learning. Its immersive learning experiences, global connectivity, personalized pathways, inclusivity, and potential for lifelong learning are reshaping our conception of education. Educators, students, policymakers, and technology developers are co-creators in this educational revolution, leveraging the Metaverse to unlock its full potential and address its challenges responsibly. As we journey further into the metaverse-enabled education landscape, it is essential to remain adaptable, innovative, and committed to the core principles of practical education. By harnessing the Metaverse's capabilities and aligning them with pedagogical goals, we can build a future where learning is not confined to classrooms but spans virtual realms, enriching lives and empowering individuals to thrive in a rapidly evolving world. The Metaverse is not a replacement for traditional education but a powerful complement, offering new dimensions and possibilities to enrich the educational journey for future generations.

In conclusion, the Metaverse emerges as a transformative force that holds boundless potential to reshape the landscape of education. Its applications in education span a broad spectrum, offering innovative solutions to longstanding challenges and ushering in a new era of learning. The Metaverse redefines how we conceive of education and empowers learners and educators through immersive experiences, global connectivity, personalized pathways, inclusivity, and lifelong learning opportunities. Immersive learning experiences, driven by virtual and augmented reality technologies, bridge the gap between theoretical knowledge and real-world application. Students can step into historical events, explore complex scientific concepts, and engage in hands-on simulations, enhancing their understanding and retention of educational content. The Metaverse's capacity for global collaboration and connectivity transcends geographical boundaries, fostering a global perspective and cultural exchange. Students from diverse backgrounds come together in virtual classrooms and shared environments, enriching their educational experiences and preparing them for a globalized world.

Personalized learning pathways, guided by artificial intelligence (AI), ensure that education meets individual needs and preferences. AI-driven adaptive platforms assess students' strengths and weaknesses, delivering tailored content and real-time feedback, optimizing the learning journey for every learner. Inclusivity takes center stage in metaverse-based education. It breaks down traditional barriers, providing accessibility features for students with disabilities and accommodating diverse learning styles. Learners from all walks of life find themselves on an equal footing, enhancing educational equity. Lifelong learning has become a reality in the Metaverse, where individuals can continuously acquire new knowledge and skills. Virtual workshops, interactive tutorials, and collaborative projects offer ongoing opportunities for personal and professional development, aligning education with the demands of a dynamic world. Educators embrace their roles as creators, designing innovative, immersive learning experiences that captivate students' imaginations and foster critical thinking. The Metaverse empowers educators to craft educational content that aligns with pedagogical goals and individual student needs. While the Metaverse offers tremendous benefits, ethical considerations are paramount. Responsible behavior, digital citizenship, and safeguards against identity theft and harassment are essential to navigating virtual spaces. In striking this transformative balance between technology and personal connection, education thrives as a vital cornerstone of personal and societal progress. The Metaverse enriches the educational journey, offering new dimensions and possibilities, while traditional educational values remain at its core. In this era of educational revolution, educators, students, policymakers, and technology developers are co-creators, collaborating to unlock the Metaverse's full potential and address its challenges responsibly. Together, they build a future where learning is not confined to classrooms but spans virtual realms, enriching lives and empowering individuals to thrive in a rapidly evolving world. The Metaverse is not a replacement for traditional education but a powerful complement, offering new horizons for future generations.

References

Nahi, A.A., Flaih, L.R., Jasim, K.F.: A perspective on the development of quantum computers and the security challenges. In: 4th International Conference on Communication Engineering and Computer Science (CIC-COCOS'2022) (2022). https://doi.org/10.24086/cocos2022/pap er.754

Abbas, S., et al.: Antecedents of trustworthiness of social commerce platforms: a case of rural communities using multi group SEM & MCDM methods. Electron. Commer. Res. Appl. **62**, 101322 (2023)

Abbate, S., Centobelli, P., Cerchione, R., Oropallo, E., Riccio, E.: A first bibliometric literature review on metaverse. In: 2022 IEEE Technology and Engineering Management Conference (TEMSCON EUROPE) (2022). https://doi.org/10.1109/temsconeurope54743.2022.9802015

Ahuja, A.S., Polascik, B.W., Doddapaneni, D., Byrnes, E.S., Sridhar, J.: The digital metaverse: applications in artificial intelligence, medical education, and integrative health. Integr. Med. Res. **12**(1), 100917 (2023). https://doi.org/10.1016/j.imr.2022.100917

Akçayır, G., Akçayır, M.: Internet use for educational purposes: university students' attitudes and opinions about copyrights. Eğitim Teknolojisi Kuram ve Uygulama **7**(1), 105 (2017). https://doi.org/10.17943/etku.288490

Al-Abrrow, H., Fayez, A.S., Abdullah, H., Khaw, K.W., Alnoor, A., Rexhepi, G.: Effect of open-mindedness and humble behavior on innovation: mediator role of learning. Int. J. Emerg. Markets (2021)

Albahri, A.S., et al.: Based on the multi-assessment model: towards a new context of combining the artificial neural network and structural equation modelling: a review. Chaos Soli-tons Fractals **153**, 111445 (2021)

AL-Fatlawey, M.H., Brias, A.K., Atiyah, A.G.: The role of strategic behavior in achievement the organizational excellence "Analytical research of the manager's views of Ur state company at Thi-Qar Governorate". J. Adm. Econ. **10**(37), 48–68 (2021)

Al-Hchaimi, A.A.J., Sulaiman, N.B., Mustafa, M.A.B., Mohtar, M.N.B., Hassan, S.L.B.M., Muhsen, Y.R.: A comprehensive evaluation approach for efficient countermeasure techniques against timing side-channel attack on MPSoC-based IoT using multi-criteria decision-making methods. Egyptian Inform. J. **24**(2), 351–364 (2023)

Alnoor, A., et al.: How positive and negative electronic word of mouth (eWOM) affects customers' intention to use social commerce? A dual-stage multi group-SEM and ANN analysis. Int. J. Hum. Comput. Interact. 1–30 (2022)

Alsalem, M.A., et al.: Rise of multiattribute decision-making in combating COVID-19: a systematic review of the state-of-the-art literature. Int. J. Intell. Syst. **37**(6), 3514–3624 (2022)

Araujo Inastrilla, C.R.: Internet search trends about the metaverse. Metaverse Basic Appl. Res. (2023). https://doi.org/10.56294/mr202326

Atiyah, A.G.: The effect of the dimensions of strategic change on organizational performance level. PalArch's J. Archaeol. Egypt/Egyptology **17**(8), 1269–1282 (2020)

Atiyah, A.G.: Strategic network and psychological contract breach: the mediating effect of role ambiguity. Int. J. Res. Manag. Stud. (IJRMS) 13(1), (2023)

Atiyah, A.G., Zaidan, R.A.: Barriers to Using Social Commerce. In Artificial Neural Networks and Structural Equation Modeling: Marketing and Consumer Research Applications, pp. 115–130. Singapore: Springer Nature Singapore (2022). https://doi.org/10.1007/978-981-19-6509-8_7

Bisset Delgado, C.: User experience (UX) in metaverse: realities and challenges. Metaverse Basic Appl. Res. **1**, 9 (2022). https://doi.org/10.56294/mr20229

Bozanic, D., Tešić, D., Puška, A., Štilić, A., Muhsen, Y.R.: Ranking challenges, risks and threats using fuzzy inference system. Decis. Making Appl. Manag. Eng. **6**(2), 933–947 (2023)

Carrisi, M.: Some considerations on the use of digital environments in learning numerical sets. In: Proceedings of the 12th International Conference on Computer Supported Education (2020). https://doi.org/10.5220/0009566504800487

Chen, Z.: Exploring the application scenarios and issues facing metaverse technology in education. interactive learning environments 1–13 (2022). https://doi.org/10.1080/10494820.2022.213 3148

Chew, X., Khaw, K.W., Alnoor, A., Ferasso, M., Al Halbusi, H., Muhsen, Y.R.: Circular economy of medical waste: novel intelligent medical waste management framework based on extension linear diophantine fuzzy FDOSM and neural network approach. Environ. Sci. Pollut. Res. 30, 1–27 (2023). https://doi.org/10.1007/s11356-023-26677-z

Contreras, G.S., González, A.H., Fernández, M.I., Cepa, C.B., Escobar, J.C.: The importance of the application of the metaverse in education. Mod. Appl. Sci. **16**(3), 34 (2022). https://doi.org/10.5539/mas.v16n3p34

Eneizan, B., Mohammed, A.G., Alnoor, A., Alabboodi, A.S., Enaizan, O.: Customer acceptance of mobile marketing in Jordan: an extended UTAUT2 model with trust and risk factors. Int. J. Eng. Bus. Manag. **11**, 1847979019889484 (2019)

Fadhil, S.S., Ismail, R., Alnoor, A.: The influence of soft skills on employability: a case study on technology industry sector in Malaysia. Interdisc. J. Inform. Knowl. Manag. **16**, 255 (2021)

Forte, M., Kurillo, G.: Cyber-archaeology and metaverse collaborative systems. Metaverse Creativity **1**(1), 7–19 (2010). https://doi.org/10.1386/mvcr.1.1.7_1

Garon, J.M.: Legal considerations for offering Metaverse-based Education. SSRN Electron. J. (2023). https://doi.org/10.2139/ssrn.4323227

Gatea, A.A., Marina, V.: Higher education funding in Iraq in terms of the experience of particular developed countries. Int. J. Adv. Stud. **6**(1), 8–17 (2016)

George Reyes, C.E.: High school students' views on the use of metaverse in mathematics learning. Metaverse **1**(2), 9 (2020). https://doi.org/10.54517/met.v1i2.1777

Germen, M.: Using 2D photography as a 3D constructional tool within the metaverse. Metaverse Creativity **1**(1), 35–50 (2010). https://doi.org/10.1386/mvcr.1.1.35_1

Ghantous, N., Fakhri, C.: Empowering metaverse through machine learning and blockchain technology: a study on machine learning, blockchain, and their combination to enhance metaverse (2022). https://doi.org/10.14293/s2199-1006.1.sor-.pp97bsj.v1

González Vallejo, R.: Metaverse, society & education. Metaverse Basic Appl. Res. (2023). https://doi.org/10.56294/mr202349

Hamid, R.A., et al.: How smart is e-tourism? A systematic review of smart tourism recommendation system applying data management. Comput. Sci. Rev. **39**, 100337 (2021)

Hashim, M.M., Kareem, M.M., Al-Azzawi, W.K., Nahi, A.A., Taha, M.S., Ali, A.H.: Based complex key cryptography: new secure image transmission method utilizing confusion and diffusion. In: 2022 8th International Conference on Contemporary Information Technology and Mathematics (ICCITM) (2022). https://doi.org/10.1109/iccitm56309.2022.10032007

Huang, H., Zeng, X., Zhao, L., Qiu, C., Wu, H., Fan, L.: Fusion of building in-formation modeling and blockchain for Metaverse: a survey. IEEE Open J. Comput. Soc. **3**, 195–207 (2022)

Hwang, G.-J., Chien, S.-Y.: Definition, roles, and potential research issues of the metaverse in education: an artificial intelligence perspective. Comput. Educ. Artif. Intell. **3**, 100082 (2022). https://doi.org/10.1016/j.caeai.2022.100082

Jameel, A.S., Rahman Ahmad, A., Alheety, A.S.: Behavioral intention to use e-wallet: the perspective of security and trust. In: 2022 2nd International Conference on Emerging Smart Technologies and Applications (eSmarTA) (2022). https://doi.org/10.1109/esmarta56775.2022.9935423

Kaddoura, S., Al Husseiny, F.: The rising trend of metaverse in education: challenges, opportunities, and ethical considerations. PeerJ Comput. Sci. **9** (2023). https://doi.org/10.7717/peerj-cs.1252

Kavut, S.: Metaverse in social life: an evaluation on metaverse news. TRT Akad-emi, **8**(17), 342–367 (2023). https://doi.org/10.37679/trta.1203028

Khaw, K.W., et al.: Modelling and evaluating trust in mobile commerce: a hybrid three stage fuzzy delphi, structural equation modeling, and neural network approach. Int. J. Hum. Comput. Interact. **38**(16), 1529–1545 (2022)

Kye, B., Han, N., Kim, E., Park, Y., Jo, S.: Educational applications of metaverse: possibilities and limitations. J. Educ. Eval. Health Prof. **18**, 32 (2021). https://doi.org/10.3352/jeehp.2021. 18.32

Lau, K.W.: Establishing a creative learning community in the immersive virtual environment for ubiquitous learning. Metaverse Creativity **5**(1), 85–102 (2015). https://doi.org/10.1386/mvcr. 5.1.85_1

Manhal, M., Al-khalidi, A., Hamad, Z.: Strategic network: Managerial myopia point of view. Management Science Letters **13**(3), 211–218 (2023)

Muhsen, Y.R., Husin, N.A., Zolkepli, M.B., Manshor, N., Al-Hchaimi, A.A.J.: Evaluation of the routing algorithms for NoC-based MPSoC: a fuzzy multi-criteria decision-making approach. IEEE Access **11**, 102806–102827 (2023)

Özdemir, A., Vural, M., Süleymanoğullari, M., Bayraktar, G.: What do university students think about the Metaverse? J. Educ. Technol. Online Learn. **5**(4), 952–962 (2022). https://doi.org/ 10.31681/jetol.1151470

Ryan, M.: Intellectual property considerations and challenges in the metaverse. SSRN Electron. J. (2023). https://doi.org/10.2139/ssrn.4484745

Sá, M.J., Serpa, S.: Metaverse as a learning environment: some considerations. Sustainability **15**(3), 2186 (2023). https://doi.org/10.3390/su15032186

Stephenson, N.: Snow crash: A novel. Spectra (2003)

Talan, T., Kalinkara, Y.: Students' opinions about the educational use of the metaverse. Int. J. Technol. Educ. Sci. **6**(2), 333–346 (2022). https://doi.org/10.46328/ijtes.385

Made Wisnu Buana, I.: Metaverse: threat or opportunity for our social world? in understanding metaverse on sociological context. J. Metaverse **3**(1), 28–33 (2023). https://doi.org/10.57019/ jmv.1144470

Zahedi, M.H., Farahani, E., Peymani, K.: A virtual e-learning environment model based on metaverse. In: 2023 10th International and the 16th National Conference on E-Learning and E-Teaching (ICeLeT) (2023). https://doi.org/10.1109/icelet58996.2023.10139894

Zhang, Q.: Rule of the Metaverse. Metaverse **3**(1), 23 (2022). https://doi.org/10.54517/met.v3i1. 1812

Zhang, X., Chen, Y., Hu, L., Wang, Y.: The Metaverse in Education: Definition, framework, features, potential applications, challenges, and future research topics. Front. Psychol. 13 (2022). https://doi.org/10.3389/fpsyg.2022.1016300

Zhang, Y., Yang, S., Hu, X., He, P.: Metaverse teaching overview. In: 2022 12th International Conference on Information Technology in Medicine and Education (ITME) (2022). https:// doi.org/10.1109/itme56794.2022.00032

Investigating Behavior of Using Metaverse by Integrating UTAUT2 and Self-efficacy

Ali Shakir Zaidan[1], Khalid Mhasan Alshammary[2], Khai Wah Khaw[1(✉)], Mushtaq Yousif[3], and XinYing Chew[4]

[1] School of Management, Universiti Sains Malaysia, 11800 Pulau Pinang, Malaysia
sha3883@student.usm.my, khaiwah@usm.my
[2] School of Distance Education, Universiti Sains Malaysia, 11800 Pulau Pinang, Malaysia
[3] Department of Accounting, Universiti Putra Malaysia, Serdang, Malaysia
[4] School of Computer Sciences, Universiti Sains Malaysia, 11800 Pulau Pinang, Malaysia
xinying@usm.my

Abstract. The Metaverse has changed the landscape of the business world to be more exciting. Metaverse technology has created a virtual reality that is similar to reality. Such technology aims to address many of the problems represented in checking virtual products, users' interaction with each other, and so on. However, the engagement and usage behavior of Metaverse suffers from hesitation by many users. This study aims to explore the influence of technology acceptance factors on Metaverse user behavior by investigating the mediation of self-efficacy. Accordingly, the current study targeted a sample of virtual network users, whose number reached 384 respondents. Structural equation modeling was used, and the results revealed that technology acceptance elements affect Metaverse user behavior. In addition, self-efficacy has fully mediated the relationship between the elements of the skill receipt model and the use of the Metaverse.

Keywords: UTAUT2 · Metaverse · Self-efficacy · Behavior intention

1 Introduction

Virtual environments have grown immensely along through the reach of the net and the popularity of community media apps. Improvements in artificial intelligence (AI) in recent years have prompted more people to relocate their everyday lives online (Gai et al., 2023). Therefore, people's time-consuming activities, such as spending, investment, and socialization, have shifted to online platforms. The word Metaverse was coined to describe a virtual world in a science fiction novel. The Metaverse is an online community where digital avatars from all around the world come together to do things like shop, work, pursue interests, go to school, and socialize (Jafar et al., 2023). The Metaverse is envisioned as a massive online virtual space where people interact through digital platforms to conduct business, find employment, and carry-on interpersonal relationships while using digital representations of themselves (avatars) in these activities. The Metaverse is a upcoming iteration of the Net in which users interact with one another using

M. Al-Emran et al. (Eds.): IMDC-IST 2024, LNNS 895, pp. 81–94, 2023.
https://doi.org/10.1007/978-3-031-51716-7_6

digital characters called avatars in a three-dimensional virtual environment. Because of how it's designed, it's possible to build online communities for purposes beyond just shopping and gaming (Roy et al., 2023). Commerce, education, employment, cultural exchange, and the forging of a new social order all seem possible in the Metaverse, which gives the impression of being a parallel reality. The Metaverse may seem like a cutting-edge technological advancement at first glance, but it's just a transformational process that shows how technology can enrich human life. Another issue with the Metaverse is that users may lose track of time while immersed in virtual reality. Individuals' connections to social media are strengthened through the usage of technology. Hence, Facebook's new suggestion to live in the metaverse is adopted, we may see this increased engagement. Snow Crash is a innovative written by Neal Stephenson in (1992). The word metaverse invented by Neal Stephenson, describes a computer-generated, multi-theoretical virtual world. Neither participants' willingness nor the factors affecting it have been assessed in prior studies (Arpaci & Bahari, 2023). The continued use of methods other than face-to-face conversation throughout the COVID-19 epidemic has been a driving force in the rise of the Metaverse. It is also proposed that the importance of factors like color, handicap, and gender be downplayed in the Metaverse. The COVID-19 epidemic also boosted consumer and business enthusiasm for digital media (Al-Emran et al., 2020).

2   Hypotheses Development

2.1   UTAUT2 Factors and Behavioral to Use Metaverse

The process of performance expectancy is described as the extent to which the use of a technology will provide consumers with benefits when performing specific tasks. Apparent helpfulness, benefit, extrinsic drive, work fit, outcome prospects and relative are the fundamental constructs underlying performance anticipation. According to the Metaverse, performance expectations were hypothesized to have a direct effect on behavioral intentions to adopt new technologies (Al-Emran et a., 2020). According to the literature review is explained what categories of targets are chosen based on the company's expectations. The company prioritizes attainable, beneficial, and realizable objectives while enhancing its performance expectancy (Gansser & Reich, 2021; Albahri et al., 2022). Businesses direct their behavior toward these objectives when they deem the expected performance and outcome to be sufficiently positive. The previous favorable or unfavorable result influences the formulation of this estimate, concept, or forecast. In other words, after a positive performance, success expectations will rise, and conversely (Lehmann et al., 2019). Prior research indicates that performance expectations are a significant factor when forecasting a desire to use a metaverse. According to Zheng et al. (2018), there is a positive correlation between performance expectations and intent to use a metaverse.

H1: Performance expectancy significantly related to the behavioral to use Metaverse.

Therefore, in this section, we discuss the effort expectancy, which is related to the ease with which customers operate technology. The main notions underpinning effort expectancy are the perceived ease of use of the metaverse and the complexity of IT (Kral et al., 2022). The Metaverse calls for a specific set of skills, knowledge, and understanding. Access to the Metaverse will increase in the future if operators trust they can usage the requisite technical equipment (including computers, mobile plans, and virtual reality headphones) with little difficulty (Lee & Kim, 2022). The UTAUT2 hypothesis postulated that the behavioral intention to use novel technology is directly affected by the anticipation of effort (Arpaci & Bahari, 2023). The grade to which a operator anticipates the metaverse to be easy to use is referred to as the effort expectations of the user. Macedo (2017) argues that the ease and convenience with which older persons use the metaverse can be inferred from their effort expectations. The ease with which people embrace metaverse services is the essence of effort expectation. Customers' expectations of how much work it will take to use IT are particularly important in the case of metaverse, which calls for specialized knowledge and abilities. Researchers such as Yang et al. (2022) discovered that students' motivation to adopt mobile learning was positively influenced by their social necessities. According to previous studies, a person's expectation of how much work will be involved in using a technology is the second most important factor in deciding whether they will utilize the tool (Carvajal et al., 2021; Manhal et al., 2023). Adoption of the metaverse will increase when the learning curve for the technology is lowered. As metaverse gets more user-friendly and less daunting, people are more possible to want to usage it. Therefore, it is important to consider the expected amount of work when advocating for the use of metaverse. As a result, making things easier can significantly impact getting more people to adopt technologies. According to Nguyen and LE (2020), outcome expectancy and effort expectancy are the major factors of entrepreneurial behavior when it comes to the usage of a technology.

H2: Effort expectancy significantly related to the behavioral to use Metaverse.

Social effect is characterized as the level to which customers consider that influential others, such as domestic, friends, and co-workers, trust they must use a specific skill, such as the metaverse app and the health care app (Oh et al., 2023; Hadi et al., 2018). Social rules and subjective norms are the fundamental concepts underlying social influence. Those social members who utilize new technologies, such as the Metaverse, may contribute to the development of confidence in the technology and provider of services (Arpaci et al., 2022). The rise in popularity of online interactions and the COVID-19 epidemic both contributed to the rapid dissemination of metaverse platforms. After going into quarantine because of the COVID-19 epidemic, people aggressively introduced immersive media tools to replace their real-world pursuits. For these and other reasons, metaverse has lately been dubbed Web 3.0 by companies like Nvidia and Meta. The network's worth grows exponentially as more users contribute to it (Mustafa et al., 2022; Krishnan et al., 2021). As a result, as the metaverse platform matures, users become more proficient in their use of it, find greater value in it, and are more pleased with their experience. Hence, social influence recognizes utility perceived simple of use, basically impacting attitudes toward metaverse and technology use (Alsalem et al., 2021; Lee & Kim, 2022). The extent of social network is correlated with the amount of supportive engagement one receives, as social influence is a relational phenomenon (Cobb, 1982; Abdullah et al., 2022; Alnoor et al., 2022). Ringle et al. (2014) argue that the quantity and quality of one's social interactions are key factors in the interchange of social influence. Having more friends in the metaverse increases the likelihood that you'll have a good time there (Oh & Yoon, 2014).

H3: Social influence significantly related to the behavioral to use Metaverse.

Facilitating conditions are defined as consumers perceptions of the incomes and care available to complete conduct (Abbas et al., 2023). The fundamental concepts original the Facilitating conditions are received behavioral control and consumer suitability (Arpaci et al., 2022). Adoption of the Metaverse requires a particular skill set and technological instruments (desktop, education platforms, handheld electronics, and VR headsets). Users would be more probable to adopt the Metaverse in the coming if they perceive they already possess the necessary technical tools and knowledge to operate in such virtual environments (Teng et al., 2022). Facilitating conditions refer to the extent to people believe a structural and technological environment happens to facilitate the usage of the metaverse. Arpaci et al. (2022), indicated that users' behavior intentions increased when users believed they could control the available resources and technology. Al-Khaldi and Wallace (1999), discovered that users' experience, knowledge, and attitude regarding personal computers influenced the likelihood that knowledge workers would use the metaverse. Using metaverse and the technology acceptance paradigm reported a positive influence of social media user facilitation conditions on the metaverse. Hence, facilitating conditions encompass the impact of essential resources, such as internet connectivity and integration with other smart devices like smartphones, as well as the requisite knowledge needed to effectively participate in health and fitness monitoring via a smartwatch (Arpaci & Bahari, 2023; Zaidan et al., 2022).

H4: Facilitating conditions significantly related to the behavioral to use Metaverse.

To what extent amusement may be obtained through information technologies, and what pleasure can be acquired from using such metaverses and technologies, is largely determined by hedonic motivation, also known as observed pleasure (Arpaci & Bahari, 2023). According to the UTAUT2, hedonic incentive has a direct and substantial impact on the likelihood that people will engage in behavior associated with using the metaverse. Previous studies have found that students' intentions to engage in metaverse-related conduct are significantly correlated with their level of hedonic motivation (Arpaci et al., 2022). Similarly, research has shown that hedonic motivation is a critical cause of the interactive purpose to utilize e-learning, and that students find metaverse applications delightful. The degree to which users take pleasure in interacting with the virtual world and cutting-edge tools is known as their hedonic motivation (Venkatesh et al., 2012). Hedonic motivation has been shown to positively correlate with the desire to participate in metaverse activities in a variety of circumstances (Farooq et al., 2017). Gao, Li, and Luo (2015) propose that individuals' propensity to adopt health care wearables is significantly affected by hedonic incentive. In the context of consumer-grade electronics, it is crucial to analyze how people intuitively interpret new technologies, as users value both practical and pleasurable features of their products (Venkatesh et al., 2016; Zaidan et al., 2022). People who enjoy using smartphones for learning and smartwatches for health monitoring are hypothesized to be more open-minded and enthusiastic about adopting and making use of the metaverse in this study (Hsu et al., 2021) (Fig. 1)).

H5: Hedonic motivation significantly related to the behavioral to use Metaverse.

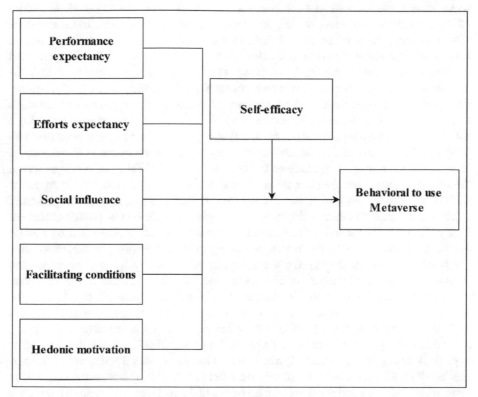

Fig. 1. Conceptual framework

2.2 Moderating Role of Self-efficacy

A person's self-efficacy is their belief in their own capacity to form and maintain relationships (Gecas, 1989). Self-efficacy was first described by Bandura, Freeman, and Lightsey (1999) as the belief that one is capable of and likely to successfully achieve goals and produce desired outcomes (Farley, 2020). The term social self-efficacy is used to characterize one's confidence in their social abilities. One definition of self-efficacy is a person's belief in his or her ability to successfully complete a given activity. Individuals who believe in their own abilities are more likely to embrace novel forms of information technology, such as the metaverse, and report higher levels of life satisfaction as a result (Guazzini et al., 2022; Abbas et al., 2021). In other words, the belief in one's own ability to quickly learn and apply cutting-edge technology has a favorable impact on the metaverse's perceived usefulness. When social media sites like YouTube, Facebook, Wikipedia, and Twitter were first introduced to the public, users' sense of self-efficacy had a role in how useful they found these sites (Oh et al., 2023). This research considers social self-efficacy as a moderator between UTAUT2 variables and Metaverse behavior. The metaverse provides fresh chances to practice and improve social skills through a wide range of interactions and problem-solving games. Interactions based on avatars on these platforms are more social, synchronous, and extensive than those based on self-exposure, praise, and sharing on earlier social media platforms (like Facebook and Instagram). The following explains the close connection between social self-efficacy and metaverse-based social support. Self-efficacy can be defined as the conviction that one can accomplish a goal because one's own efforts alone (Alhwaiti, 2023). Social competence can be improved by increasing exposure to opportunities to practice social skills and engage in constructive social interactions (Atiyah, 2020). The internet provides a haven for people to experiment with new social circumstances and build connections with people from all walks of life. Second, unlike text-based asynchronous communication, users in the metaverse can participate in vocal and nonverbal forms of communication through their avatar that are more primitive and global (Salimon et al., 2023; Atiyah et a., 2022). Avatar-based interaction can improve students' feelings of social presence, which in turn improves their ability to learn. People can learn how to make the most of their offline social connections and create a supportive social environment by modeling these interactions in the metaverse, practicing them, and building on their successes (Trautner & Schwinger, 2020; AL-Fatlawey et al., 2021). Finally, human beings develop their competency in a range of social circumstances and new interactions as they gain experience interacting with others who have varied points of view (Gecas, 1989). The metaverse's virtual worlds are a great way to meet people from all walks of life and expand your social circle, all while having fun and learning something new. Having more friends in your social circle is linked to better social skills, according to studies (e.g., Valkenburg & Peter, 2008). It was found to have a beneficial influence on the perception of comfort of use. Furthermore, Rafiola et al. (2020), validates the connection between self-efficacy and the sense of ease of use in the metaverse, finding that people's confidence in their own abilities has a beneficial impact on this estimation.

H6: The relationship between UTAUT2 factors and behavioral to use Metaverse would be stronger when self-efficacy is high.

3 Methodology

3.1 Data Collection

The current study's population is Malaysian users. The target respondents were users of virtual networks in Malaysian. Google Forms was used to collect 384 questionnaires of respondents. The G*Power 3 (Faul et al., 2007) was used to regulate the lowest sample scope with an effect size of 0.15, a margin of error of 5%, and five predictors. Based on this approach, the minimum is 92; therefore, the sample size in the current study exceeded the required size. The questionnaire included 49 questions, in which respondents were to degree their answers based on a 5-point Likert Scale (1 = "strongly disagree" to 5 = "strongly agree").

3.2 Measurement of Variables

The variables in the research model were measured in this study using established scales from the body of existing literature. A three-item scale was used to measure performance expectations, which was created by Arpaci et al. (2023). In contrast, the effort expectancy scale consisted of four items, while a three-item scale was used to assess social influence. Moreover, a four-item scale was used to measure facilitating conditions and a three-item scale to measure hedonic motivation. A three-item scale was used to measure Metaverse and the same number of items scale to measure self-efficacy that established by Sheaffer et al. (2011).

3.3 Common Method Bias

The same scale approach was utilized to measure the study variables. Therefore, this may lead to the problem of common method bias, which happens when there is systematic variation in responses due to the employment of a common scaling strategy on measures drawn from the same data source (Fuller et al., 2016). In order to identify prevalent technique bias, the authors used Harman's one-factor test (Maxwell & Harman, 1968). Harman's one-factor test revealed a lower-than-ideal 27% total variation for a single component. Indicating that the usual technique variance did not influence the results.

4 Data Analysis

Descriptive analysis in this study was conducted by SPSS version 26. Partial Least Squares Structural Equation Modelling (PLS-SEM) was employed to validate the research model by SmartPLS 4 (Ringle et al., 2015). The following justify the use of PLS-SEM in this investigation: using dormant variable star scores in an examination of prognostic relevance (Hair et al., 2017); (1) focusing on the prediction of the endogenous variables; (2) research model (expressed by the hypotheses) is extremely complex. The statistical analysis includes two phases: the assessment measurement of model and the assessment physical model.

4.1 Assessment of Measurement Model

The measurement model was conducted to ensure the validity and reliability of the model. It includes convergent and discriminant validity. Convergence validity was evaluated using factor loadings, composite reliability (CR), and average variance extracted (AVE) (Hair et al., 2017). All item loadings were greater than 0.5, AVE values were greater than 0.5, and CR was greater than 0.7 (Hair et al., 2017), indicating that the measurement model met content validity requirements, as exposed in Table 1.

Table 1. Results of the Measurement Model.

Items	FL	CR	AVE
	>0.7	>0.7	>0.5
Behavioral to use Metaverse		0.92	0.792
BUM1	0.881		
BUM2	0.919		
BUM3	0.87		
Performance Expectancy		0.736	0.736
PE1	0.876		
PE2	0.815		
PE3	0.882		
Efforts Expectancy		0.879	0.645
EE1	0.78		
EE2	0.766		
EE3	0.815		
EE4	0.849		
Social Influence		0.889	0.729
SI1	0.802		
SI2	0.903		
SI3	0.853		
Facilitation Conditions		0.915	0.729
FC1	0.867		
FC2	0.864		
FC3	0.824		
FC4	0.858		
Hedonic Motivation		0.896	0.742
HM1	0.824		
HM2	0.866		
HM3	0.892		
Self-efficacy		0.903	0.756
SE1	0.871		
SE2	0.868		

(continued)

Table 1. (*continued*)

Items	FL	CR	AVE
	>0.7	>0.7	>0.5
SE3	0.87		

Abbreviations: BUM, Behavioral to use Metaverse; PE, Performance Expectancy; EE, Efforts Expectancy; SI, Social Influence; FC, Facilitation Conditions; HM, Hedonic Motivation; SE, Self-efficacy

According to Henseler et al. (2015), the heterotrait-monotrait ratio of correlations (HTMT) is a more reliable approach; therefore, it was used to evaluate discriminant validity. All the discriminant values did not exceed the HTMT.85 (Kline, 2015) criterion. Thus, the variables items have no multi-collinearity issues (see Table 2).

Table 2. Discriminant Validity Based on HTMT Method

Constructs	1	2	3	4	5	6	7
1. Efforts Expectancy							
2. Behavioral to use Metaverse	0.765						
3. Facilitation Conditions	0.802	0.624					
4. Hedonic Motivation	0.752	0.604	0.833				
5. Performance Expectancy	0.809	0.876	0.795	0.802			
6. Self-efficacy	0.829	0.801	0.823	0.828	0.744		
7. Social Influence	0.821	0.542	0.845	0.829	0.718	0.781	

4.2 Assessment of Structural Model

The structural model aims to explain the causal relationships between the exogenous and endogenous variables. After obtained the desired level of validity for the dimension model, the physical model was assessed in the second phase of analysis. The number of resamples used in this bootstrapping exercise was 5000 by using SmartPLS. Path analysis was used to verify the study's hypotheses. Inner VIF values, path-coefficient results, coefficient of determination (R^2), effect sizes (f^2), and predictive relevance Q^2 have been presented, as recommended by (Hair. et al. 2017). There was a requirement to ensure that collinearity subjects did not contaminate the reversion consequences, given that the PLS-SEM calculation of the track constants connecting the concepts was based on a sequence of reversion examines. All VIF inner values were less than 5, indicating that collinearity is not a serious problem across all of the methods, as shown in Table 3.

The results of the significance of the proposed path indicate that performance expectancy ($\beta = 0.481$, t = 10.021, p < .001), efforts expectancy ($\beta = 0.192$, t =

Table 3. Structural Model and Hypotheses Testing

	Path coefficient	St. Beta	Mean	St. Error	T-t	P-v	2.50% LB	97.50% UB	Result
H1	PE - > BUM	0.481	0.483	0.048	10.021	0.000	0.074	0.312	Accept
H2	EE - > BUM	0.192	0.193	0.061	3.125	0.001	0.047	0.159	Accept
H3	SI - > BUM	0.090	0.089	0.041	2.209	0.014	−0.204	−0.012	Accept
H4	FC - > BUM	0.514	0.513	0.055	9.408	0.000	0.404	0.619	Accept
H5	HM - > BUM	0.111	0.112	0.049	2.271	0.012	0.101	0.298	Accept
H6	SE x HM- > BUM	0.055	0.055	0.026	2.110	0.035	0.004	0.107	Accept
H7	SE x PE - > BUM	0.191	0.190	0.041	4.644	0.000	0.154	0.031	Accept
H8	SE x FC - > BUM	0.159	0.161	0.044	3.628	0.000	−0.273	−0.110	Accept
H9	SE x EE - > BUM	0.138	0.138	0.038	3.608	0.000	0.213	0.061	Accept
H10	SE x SI - > BUM	0.104	0.106	0.038	2.725	0.003	0.133	0.085	Accept

3.125, p < .001), social influence ($\beta = 0.090$, t = 2.209, p < .001), facilitation conditions ($\beta = 0.514$, t = 9.408, p < .001), and hedonic motivation ($\beta = 0.111$, t = 2.271, p < .001) are positively related to behavioral to use metaverse (see Table 3). Therefore, hypotheses H1, H2, H3, H4, and H5 have been accepted. Furthermore, moderation analysis results refer to that self-efficacy is found to significantly moderate the association among performance expectation ($\beta = 0.,191$ t = 4.644, p < .001), efforts expectancy ($\beta = 0.138$, t = 3.608, p < .001), social influence ($\beta = 0.104$, t = 2.725, p < .001), facilitation conditions ($\beta = 0.519$, t = 3.628, p < .001), and hedonic motivation ($\beta = 0.055$, t = 2.110, p < .001) and behavioral to use metaverse (see Table 2). As a result, the moderation hypothesis H6, H7, H8, H9, H10, are supported.

The researchers rated R^2 between 0 and 1, the larger values referring to high prediction accuracy (Appannan et al., 2022). According to Chin (1998), R^2 value of 0.67, 0.33, and 0.19 in PLS path models are considered high, moderate, and low, respectively. The result shows that the value of R in the present study refers to high predictive accuracy. Path model significance of prediction for a given reliant on construct is designated by a Q^2 value larger than zero for a specific reflective endogenic latent factor (Hair et al., 2010). Q^2 value ($Q^2 = 0.34$) is greater than zero in Table 3. This means the model is useful for making predictions. The significance of an exogenic construct's effect on an endogenous concept is described by its effect size (f^2) value (Cohen, 1988). According to Cohen (1988), effect size values of 0.02, 0.15, and 0.35 classify as small, medium, and large effects, respectively. The f^2 values in the present study ranged between (0.197 to 0.041), which refers to a medium effect.

5   Conclusion

Several gaps in the existing literature are filled by the current investigation. At the outset, it has been established that UTAUT2 factors are linked to the confidence one feels when using the Metaverse. To fill this void, we incorporated UTAUT2 variables and self-efficacy into our model of Metaverse usage behavior. Important research should be conducted to ensure that these factors were not selected at random but rather after a comprehensive examination of the literature on the behavioral motivations to participate in the Metaverse. Second, this study related the major characteristics of UTAUT2 with self-efficacy, demonstrating the model's efficacy in explaining metaverse, and adding to the rising body of works on the model's application in the behavioral to use Metaverse adoption area. Furthermore, self-efficacy explains more variation (90%) than behavior (88%) when it comes to using the Metaverse in UTAUT2. Finally, the empirical data analysis provides strong support for the hypothesized correlations between self-efficacy and factors like presentation expectation, effort expectation, enabling conditions, hedonistic incentive, and social effect. This finding calls for more research on the current state and defining characteristics of Metaverse usage, especially in less developed nations like Malaysia. Therefore, decision-makers who place a strong emphasis on using the Metaverse in their organizations must offer training sessions to their users to help them apply for it. To encourage the long-term sustainable usage of the Metaverse environment, designers and developers must simultaneously enhance the operator sociability of the situation, particularly resolve difficulties. Institutions, organizations, and commercial enterprises using the Metaverse need to consider the enabling circumstances supporting its social sustainability. Because the requisite levels of knowledge vary from one area to another, these conditions do as well. These organizations must be able to pay for the equipment (such VR headsets) and software required to access the Metaverse. To exploit these new virtual environments, users also need to be equipped with specialized VR and AR information and knowhow. The avatars that users use to represent themselves in this virtual environment should be manageable for users. Organizations need to have set tactics on how the Metaverse is connected to their long-standing objectives if they are to reaffirm individuals' commitment to being conscientious. In light of this, people will try to leverage these new surroundings to keep their jobs, especially given the government's push for a digital society. Users typically develop their extraversion abilities through social contact. Although the Metaverse encourages user participation, this has no impact on how socially sustainable it is (Albahri et al., 2022). To meet the needs of various institutions in the future, people are urged to experiment with using the Metaverse in their everyday lives and experience the taste of connection with their networks and coworkers.

References

Abbas, S., Al-Abrrow, H., Abdullah, H.O., Alnoor, A., Khattak, Z.Z., Khaw, K.W.: Encountering Covid-19 and perceived stress and the role of a health climate among medical workers. Current Psychol. 1-14 (2021)

Abbas, S., et al.: Antecedents of trustworthiness of social commerce platforms: a case of rural communities using multi group SEM & MCDM methods. Electron. Commer. Res. Appl. **62**, 101322 (2023)

Abdullah, H.O., Atshan, N., Al-Abrrow, H., Alnoor, A., Valeri, M., Erkol Bayram, G.: Leadership styles and sustainable organizational energy in family business: modeling non-compensatory and nonlinear relationships. J. Fam. Bus. Manag. **13**, 1104–1131 (2022)

Albahri, O.S., et al.: Novel dynamic fuzzy decision-making framework for COVID-19 vaccine dose recipients. J. Adv. Res. **37**, 147–168 (2022)

Al-Emran, M., Al-Maroof, R., Al-Sharafi, M.A., Arpaci, I.: What impacts learning with wearables? An integrated theoretical model. Interact. Learn. Environ. **30**(10), 1897–1917 (2022)

AL-Fatlawey, M.H., Brias, A.K., Atiyah, A.G.: The role of Strategic Behavior in achievement the Organizational Excellence "Analytical research of the manager's views of Ur State Company at Thi-Qar Governorate". J. Adm. Econ. **10**(37), 48–68 (2021)

Alhwaiti, M.: Acceptance of artificial intelligence application in the post-covid era and its impact on faculty members' occupational well-being and teaching self efficacy: a path analysis using the UTAUT 2 model. Appl. Artif. Intell. **37**(1), 2175110 (2023)

Al-Khaldi, M.A., Wallace, R.O.: The influence of attitudes on personal computer utilization among knowledge workers: the case of Saudi Arabia. Inform. Manag. **36**(4), 185–204 (1999)

Alnoor, A., et al.: How positive and negative electronic word of mouth (eWOM) affects customers' intention to use social commerce? A dual-stage multi group-SEM and ANN analysis. Int. J. Hum. Comput. Interact. 1–30. (2022). https://doi.org/10.1080/10447318.2022.2125610

Alsalem, M.A., et al.: Based on T-spherical fuzzy environment: a combination of FWZIC and FDOSM for prioritising COVID-19 vaccine dose recipients. J. Infect. Public Health **14**(10), 1513–1559 (2021)

Appannan, J.S., Mohd Said, R., Ong, T.S., Senik, R.: Promoting sustainable development through strategies, environmental management accounting and environmental performance. Bus. Strategy Environ. March, 1–17 (2022). https://doi.org/10.1002/bse.3227

Arpaci, I., Bahari, M.: Investigating the role of psychological needs in predicting the educational sustainability of metaverse using a deep learning-based hybrid SEM-ANN technique. Interact. Learn. Environ. 1–13 (2023). https://doi.org/10.1080/10494820.2022.2164313

Arpaci, I., Karatas, K., Kusci, I., Al-Emran, M.: Understanding the social sustainability of the metaverse by integrating UTAUT2 and big five personality traits: a hybrid SEM-ANN approach. Technol. Soc. **71**, 102120 (2022)

Atiyah, A.G.: The effect of the dimensions of strategic change on organizational performance level. PalArch's J. Archaeol. Egypt/Egyptology **17**(8), 1269–1282 (2020)

Carvajal-Trujillo, E., Molinillo, S., Liébana-Cabanillas, F.: Determinants and risks of intentions to use mobile applications in museums: An application of fsQCA. Curr. Issue Tour. **24**(9), 1284–1303 (2021)

Chin, W.W.: Issues and opinion on structural equation modeling. MIS Quarterly: Manag. Inform. Syst. **22**(1), vii–xvi. JSTOR (1998)

Cobb, A.T.: A social psychological approach to coalition membership: An expectancy model of individual choice. Group Organ. Stud. **7**(3), 295–319 (1982)

Cohen, J.: Statistical power Analysis for the Behavioral Sciences, 2nd edn. (1988). https://doi.org/10.4324/9780203771587

Farley, H.: Promoting self-efficacy in patients with chronic disease beyond traditional education: a literature review. Nurs. Open **7**(1), 30–41 (2020)

Farooq, M.S., et al.: Acceptance and use of lecture capture system (LCS) in executive business studies: extending UTAUT2. Interact. Technol. Smart Educ. **14**(4), 329–348 (2017)

Faul, F., Erdfelder, E., Lang, A.G., Buchner, A.: G*Power 3: a flexible statistical power analysis program for the social, behavioral, and biomedical sciences. Behav. Res. Methods **39**(2), 175–191 (2007). https://doi.org/10.3758/BF03193146

Fuller, C.M., Simmering, M.J., Atinc, G., Atinc, Y., Babin, B.J.: Common methods variance detection in business research. J. Bus. Res. **69**(8), 3192–3198 (2016). https://doi.org/10.1016/j.jbusres.2015.12.008

Gai, T., Wu, J., Cao, M., Ji, F., Sun, Q., Zhou, M.: Trust chain driven bidirectional feedback mechanism in social network group decision making and its application in metaverse virtual community. Expert Syst. Appl. **228**, 120369 (2023)

Gansser, O.A., Reich, C.S.: A new acceptance model for artificial intelligence with extensions to UTAUT2: an empirical study in three segments of application. Technol. Soc. **65**, 101535 (2021)

Gecas, V.: The social psychology of self-efficacy. Ann. Rev. Sociol. **15**(1), 291–316 (1989)

Guazzini, A., Serritella, E., La Gamma, M., Duradoni, M.: Italian version of the internet self-efficacy scale: Internal and external validation. Hum. Behav. Emerg. Technol. **2022**, 9347172 (2022)

Hadi, A.A., Alnoor, A., Abdullah, H.O.: Socio-technical approach, decision-making environment, and sustainable performance: Role of ERP systems. Interdiscip. J. Inf. Knowl. Manag. **13**, 397–415 (2018)

Hair, F.J., Hult, G.T., Ringle, C., Sarstedt, M.: A Primer on Partial Least Squares Structural Equation Modeling (PLS-SEM), Sage (2017)

Hair, J., Black, W., Babin, B., Anderson, R.: Multivariate Data Analysis: A Glob-al Perspective, 7th edn. (2010)

Hsu, S.H.Y., Tsou, H.T., Chen, J.S.: "Yes, we do. Why not use augmented reality?" customer responses to experiential presentations of AR-based applications. J. Retail. Consum. Serv. **62**, 102649 (2021)

Jafar, R.M.S., Ahmad, W., Sun, Y.: Unfolding the impacts of metaverse aspects on telepresence, product knowledge, and purchase intentions in the metaverse stores. Technol. Soc. **74**, 102265 (2023)

Kline, R.B.: Principles and Practice of Structural Equation Modeling. Guilford Publications (2015)

Kral, P., Janoskova, K., Potcovaru, A.M.: Digital consumer engagement on blockchain-based metaverse platforms: extended reality technologies, spatial analytics, and immersive multisensory virtual spaces. Linguist. Philos. Inv. **21**, 252–267 (2022)

Krishnan, E., et al.: Interval type 2 trapezoidal-fuzzy weighted with zero inconsistency combined with VIKOR for evaluating smart e-tourism applications. Int. J. In-tell. Syst. **36**(9), 4723–4774 (2021)

Lee, U.K., Kim, H.: UTAUT in metaverse: an "Ifland" case. J. Theor. Appl. Electron. Commer. Res. **17**(2), 613–635 (2022)

Lehmann, A.I., Brauchli, R., Bauer, G.F.: Goal pursuit in organizational health interventions: the role of team climate, outcome expectancy, and implementation intentions. Front. Psychol. **10**, 154 (2019)

Macedo, I.M.: Predicting the acceptance and use of information and communication technology by older adults: An empirical examination of the revised UTAUT2. Comput. Hum. Behav. **75**, 935–948 (2017)

Manhal, M., Al-khalidi, A., Hamad, Z.: Strategic network: managerial myopia point of view. Manag. Sci. Lett. **13**(3), 211–218 (2023)

Maxwell, A.E., Harman, H.H.: Modern Factor Analysis. J. Roy. Stat. Soc. A (General) **131** (1968)

Mustafa, S., Zhang, W., Anwar, S., Jamil, K., Rana, S.: An integrated model of UTAUT2 to understand consumers' 5G technology acceptance using SEM-ANN approach. Sci. Rep. **12**(1), 20056 (2022)

Nguyen, H.Q., Le, O.T.T.: Factors affecting the intention to apply management accounting in enterprises in Vietnam. J. Asian Finance, Econ. Bus. (JAFEB) **7**(6), 95–107 (2020)

Oh, H.J., Kim, J., Chang, J.J., Park, N., Lee, S.: Social benefits of living in the metaverse: the relationships among social presence, supportive interaction, social self-efficacy, and feelings of loneliness. Comput. Hum. Behav. **139**, 107498 (2023)

Oh, J.C., Yoon, S.J.: Predicting the use of online information services based on a modified UTAUT model. Behav. Inform. Technol. **33**(7), 716–729 (2014)

Rafiola, R., Setyosari, P., Radjah, C., Ramli, M.: The effect of learning motivation, self-efficacy, and blended learning on students' achievement in the industrial revolution 4.0. Int. J. Emerg. Technol. Learn. (iJET) **15**(8), 71–82 (2020)

Roy, R., Babakerkhell, M.D., Mukherjee, S., Pal, D., Funilkul, S.: Development of a framework for metaverse in education: a systematic literature review approach. IEEE Access **11**, 57717–57734 (2023)

Ringle, C.M., da Silva, D., Bido, D.d.S.: Structural equation modeling with the smartpls. ReMark - Revista Brasileira De Marketing **13**(2), 56–73 (2014). https://doi.org/10.5585/remark.v13i2.2717

Salimon, M.G., Kareem, O., Mokhtar, S.S.M., Aliyu, O.A., Bamgbade, J.A., Adeleke, A.Q.: Malaysian SMEs m-commerce adoption: TAM 3, UTAUT 2 and TOE approach. J. Sci. Technol. Policy Manag. **14**(1), 98–126 (2023)

Teng, Z., Cai, Y., Gao, Y., Zhang, X., Li, X.: Factors affecting learners' adoption of an educational metaverse platform: an empirical study based on an extended UTAUT model. Mobile Inform. Syst. **2022**, 1–15 (2022)

Trautner, M., Schwinger, M.: Integrating the concepts of self-efficacy and motivation regulation: how do self-efficacy beliefs for motivation regulation influence self-regulatory success? Learn. Individ. Differ. **80**, 101890 (2020)

Valkenburg, P.M., Peter, J.: Adolescents' identity experiments on the internet: consequences for social competence and self-concept unity. Commun. Res. **35**(2), 208–231 (2008)

Venkatesh, V., Bala, H., Sambamurthy, V.: Implementation of an information and communication technology in a developing country: a multimethod longitudinal study in a bank in India. Inf. Syst. Res. **27**(3), 558–579 (2016)

Venkatesh, V., Thong, J.Y., Xu, X.: Consumer acceptance and use of information technology: extending the unified theory of acceptance and use of technology. MIS Q. **36**(1), 157–178 (2012)

Yang, F., Ren, L., Gu, C.: A study of college students' intention to use metaverse technology for basketball learning based on UTAUT2. Heliyon **8**(9), e10562 (2022)

Zaidan, A.S., Chew, X., Khaw, K.W., Ferasso, M.: Electronic word of mouth and social commerce. In: Alnoor, A., Wah, K.K., Hassan, A. (eds.) Artificial neural networks and structural equation modeling: marketing and consumer research applications. Singapore: Springer Nature Singapore, pp. 79–95 (2022). https://doi.org/10.1007/978-981-19-6509-8_5

Understating the Social Sustainability of Metaverse by Integrating Adoption Properties with Users' Satisfaction

Abbas Gatea Atiyah[1], Nagham Dayekh Abd All[2], Ali Shakir Zaidan[3], and Gül Erkol Bayram[4(✉)]

[1] College of Administration and Economic, University of Thi-Qar, Nasiriyah, Iraq
`abbas-al-khalidi@utq.edu.iq`
[2] College of Administration and Economic, Karbala University, Karbala, Iraq
`nagham.d@uokerbala.edu.iq`
[3] School of Management, Universiti Sains Malaysia, 11800 Pulau Pinang, Malaysia
`sha3883@student.usm.my`
[4] School of Tourism and Hospitality Management Department of Tour Guiding, Sinop University, Sinop, Turkey
`gulerkol@windowslive.com`

Abstract. Metaverse is a technical tool that captures the vast virtual world. It helps users interact and discover information through symbolic images and diagrams. Therefore, metaverse greatly expands the process of interaction between users from different regions. It is important to answer how achieving the social sustainability for this tool. This study aimed to build a conceptual model to achieve the social sustainability of the metaverse. Depending on a set of important elements that affect the user's satisfaction. Seven hypotheses have been postulated among the factors mentioned. The results showed that user satisfaction is a key element in achieving the prediction of social sustainability. The study concluded that the sustainability of metaverse lies in user satisfaction.

Keywords: Social Sustainability · Metaverse · Perceived Trialability · Perceived Observability · Perceived Compatibility · User's Satisfaction

1 Introduction

Metaverse is one of the most capable technologies for realizing virtual reality (Hankookilbo 2022). This term Metaverse was first circulated in a science fiction novel. The purpose of using it was to describe a wide and three-dimensional virtual environment (Hwang and Lee 2022). In the development of its application process, it has achieved human interaction through advanced technology and with the help of the Internet. Therefore, many platforms and applications are trying to take advantage of it (Suzuki et al. 2020). For example, in the areas of education and learning, in the work of airlines, and other businesses. In its work, Metaverse relies on creating a computerized infrastructure in every workable and tradable place (Tang 2021; Erdem and Sekar 2022). Users can use

M. Al-Emran et al. (Eds.): IMDC-IST 2024, LNNS 895, pp. 95–107, 2023.
https://doi.org/10.1007/978-3-031-51716-7_7

various metaverse applications via mobile phone, PC and other portable devices (Lee et al. 2021). As a result of the importance of the metaverse in virtual reality, studies have been conducted in the field of its mobile applications. Some of which emphasized the importance of analyzing the reactions of metaverse users. Because of their role in sustaining the success of metaverse applications in the future (Yavuz et al. 2021). Where some of these studies showed that the analysis of user feedback lies in the level of satisfaction with metaverse services (Jang 2017; Alameri et al. 2019; Isaac et al. 2018; Bilgihan et al. 2018; Gatea and Marina 2016). On the other hand, some studies have linked user satisfaction with the ability to predict the extent to which they will use metaverse services in the future, and what resources are needed to re-use them for the same applications (Wang et al. 2021; Choi et al. 2021; Lee et al. 2022; Eneizan et al. 2019). This indicates the importance of discussing the sustainability of metaverse services and the need for them to be effective and exist without obstacles that limit the possibility of their use. However, there is a rarity of studies on the sustainability of metaverse applications. Although it is of a social nature and is greatly affected by the reactions of users. Therefore, its social sustainability must be examined. Because the metaverse is based primarily on technology and is therefore an element that can be developed and corrected. There is a possibility to increase or decrease resources and capabilities in order to achieve the possibility of survival and continuity of its services and to adapt to the necessary requirements of the user (Alnoor et al. 2022; Al-Abrrow et al. 2021). Based on this, a conceptual model was built aiming at the social sustainability of the metaverse, despite what the researchers pointed out in their study of the metaverse of the huge data and their lack of reliance on traditional methodologies in analysis such as the Structural Equation Model (SEM) or the Technology Acceptance Model (TAM) (Lee et al. 2022; Khaw et al. 2022; Alsalem et al. 2022). Our study adopted a more accurate analysis in response to the requirements of the magnitude of the data. The study reached a set of conclusions. It shows that there is an effect of adoption properties elements in determining the level of user satisfaction, and therefore user satisfaction clearly affects the achievement of social sustainability for the metaverse (Albahri et al. 2021). Also, user satisfaction significantly mediates the relationship between adoption properties components and metaverse sustainability.

2   Hypotheses Development

2.1   Perceived Trialability and User's Satisfaction

The manufacturer always aspires to achieve the maximum degree of ease in knowing the innovation points of his products (AL-Fatlawey et al. 2021). Therefore, adopts perceived trialability as a tool that has a role in facilitating the discovery of ideas that the manufacturer puts in the product or in the method of marketing it (Manhal et al. 2023). Which facilitates the try it at no cost to the user (Lee et al. 2011; Martins et al. 2004; Sonnenwald et al. 2001; Atiyah 2023). This method helps the user to have prior information about the product. And thus, reduce errors in the use of the product in the future or the method of preservation (Lee 2007). This depends on the nature of this product, such as an electrical device or a type of food…etc. As there is no doubt that the innovation proposed by one of the manufacturers aims through it to reduce dependence on a product present in the market, and therefore it aims to highlight its innovative product and prove its value, by achieving user satisfaction (Johnson et al. 2018; Atiyah 2023). Because user feedback represents the outcome of all efforts made by the manufacturer. Therefore, modern opinions announce the importance of perceived trialability in achieving consumer satisfaction (Akour et al. 2022). Thus, we assume:

H1: Perceived trialability would be positively link with user's satisfaction.

2.2 Perceived Observability and User's Satisfaction

Companies closely monitor people's reactions to their products. Especially if those products were just offered, as the first mover (Wheelen et al. 2018). To the same extent, or perhaps to a greater extent, companies monitor feedback on the developed product. The reason is the abundance of competition among companies, which leads to a move towards innovation in the product or service accompanying it, such as shipping or after-sales services (Porter and Heppelmann 2015). What is the target of developing the product? Certainly, to achieve user satisfaction. So, what is the best tool through which the company draws the attention of the consumer to its products? Certainly, perceived observability. Therefore, companies are working hard to raise the level of observation among individuals who view their products (Johnson et al. 2018). Studies have shown that the possibility of observation is positively related to the level of effort made by companies to achieve it realistically (Al-Gahtani 2003). Companies that don't try hard enough cannot successfully employ their innovations and thus fail to achieve the effectiveness of observation. So, it can be assumed:

H2: Perceived observability would be positively link with user's satisfaction.

2.3 Perceived Compatibility and User's Satisfaction

One of the most important basic elements in achieving perceived compatibility is that companies before they start developing the product or related services, conduct surveys through focus groups (which are used recently) to identify what the customer desires or what exceeds his ambitions (Bruseberg and Deana 2002). Based on the results achieved, companies work to develop the product in a way that achieves the perceived compatibility with psychological and cultural standards and previous knowledge (Greenhalgh et al. 2004; Alnoor et al. 2022). Technology has become the main element in making these developments, especially with the expansion of competition and the possibility of rapid access to products, and the invasion of competitors' markets (Chew et al. 2023; Abbas et al. 2023). Therefore, companies began to search for how to use technology to lead to user satisfaction and happiness (Akour et al. 2022). This is done by using technology in developing the product and making it compatible with the standards in the mind of the customer and connecting that development to the previous experiences of the user (Rogers 2003). Previous studies have shown that perceived compatibility provides an important incentive for the purchase intention of the potential consumer and gives a greater possibility of building a positive image among individuals about the developed product (Aubert and Hamel 2001; Denis et al. 2002; Akour et al. 2022; Hamid et al. 2021). So, we assume:

H3: Perceived compatibility would be positively link with user's satisfaction.

2.4 Perceived Complexity and User's Satisfaction

When the manufacturer cannot properly translate his ideas, manufacturer will made or develop products with complex technology. The perceived complexity is the difficulty in understanding the technology by potential consumers (Hardgrave et al. 2003; Atiyah 2023). The literature shows that the higher the perceived complexity, led to lower user's experience intention (Akour et al. 2022). So, we find there is an emphasis on the importance of reducing perceived complexity. Especially since technology has come to reduce complexity and facilitate the process of using products (Tobbin 2010; Atiyah 2023). Because it is not easy to use technology by individuals. This requires a great effort of companies in employing technology in a way that makes it easier for the user to use the product (Shih 2007; Manhal et al. 2023). Where many companies have followed usage simulation systems and launched video experiments and vital applications that show the process of successful use of the developed product. For example, Apple recently launched a new feature when purchasing a newer version of the iPhone. It is the ability to synchronize between the old and developed devices, which leads to transferring all data to the developed device without the slightest effort. This reduces the perceived complexity of using the developed product and thus achieves user satisfaction. This is consistent with (Akour et al. 2022; Atiyah 2023) that companies should use technology in a way that reduces complexity, not the other way around. So, we assume the following:

H4: Perceived complexity would be positively link with user's satisfaction.

2.5 Personal Innovativeness and User's Satisfaction

When an individual acts in an innovative manner in understanding the technology embedded in the product. Individual shows a kind of willingness to adopt the product (Khan et al., 2019). The degree of willingness by users relates to their level of acceptance of the technology (Agarwal and Prasad 1998; Atiyah 2020). In other words, personal innovation refers to the individual's readiness and acceptance of using the developed product. Therefore, readiness is an integral part of personal innovation, as an external factor for measuring user acceptance of technology (Khan and Ullah 2014; Atiyah 2023). Some researchers have studied the importance of personal innovation and its relationship to user satisfaction. Where they explained that there is an impact of personal innovation in the crystallization of user satisfaction (Khan et al. 2019; Akour et al. 2022). Innovation demonstrates the company's ability to provide products that can be analyzed, measured, imagined, and innovative in use. As well as the ability to repair and rehabilitate. These elements are of importance to the user in allocating his ability to a type of developed product that can be acquired. Therefore, the following can be assumed:

H5: Personal innovativeness would be positively link with user's satisfaction.

2.6 User's Satisfaction and Social Sustainability of the Metaverse

In the year (2022) at the Consumer Electronics Show (CES), the metaverse was evaluated as one of the most important future technologies (Hankookilbo 2022; Atiyah 2022). It adopts technology as a key element in achieving accessibility, as well as other services. As other platforms and applications increasingly seek to take advantage of the technology used by the metaverse (Suzuki et al. 2022). Today, it is involved in many critical and important areas of life, such as booking airline tickets, distance learning, as well as various applications on mobile phones or other tablets. Therefore, it became necessary to search for how to achieve sustainability for this tool. And since dealing through the metaverse provides a social environment that fosters different classes of society, different ages, different tastes, and different experiences (Alfaisal et al. 2022). Therefore, we must look for its social sustainability. By social sustainability we mean the continuity of broad acceptance and satisfaction with the services provided across the metaverse. The continuity of user satisfaction is evident in the increase and continuity of positive dialogues, encouraging feedback, and the possibility of increasing the positive approvals shown by users. What is also encouraging to achieve social sustainability is the significant increase in the number of metaverse applications using mobile devices, due to the advantage of successfully employing technology. As well as satisfaction in the use of this tool. This is what the study concluded (Lee et al., 2022). Therefore, it can be assumed:

H6: user's satisfaction would be positively link with social sustainability of the Metaverse.

2.7 Mediating Role of User's Satisfaction

Metaverse sustainability is of paramount importance. It is one of the tools that has achieved widespread success (Lee et al., 2022). As it is necessary to search for certain aspects that help in achieving sustainability. One of the most important aspects of achieving sustainability is the user. The user is the most vital element in the metaverse's sustainability chain because user feedback represents the sensory and emotional aspect of the level of services that companies bring to market (Akour et al. 2022; AL-Fatlawey et al. 2021). Therefore, the researchers were interested in how to achieve user satisfaction, but the matter extended beyond that, which is achieving user happiness. Certainly, this reflects in the metaverse, as it represents the reader's portal through which looks at a wide horizon of user reactions (Jang and Yi 2017; Yin et al., 2022; Atiyah 2020). Thus, it is supported by a large number of opinions, dialogues, and discussions that indicate its use and the indispensability of it. So, it's safe to say that the user plays a vital role in the sustainability of the metaverse. Thus, we can assume:

H7: There is a mediating role of user's satisfaction in relationship between adoption properties and social sustainability of the metaverse.

3   Research Methodology

A survey method was used in this study. More specifically, a questionnaire technique was used to gather primary data that was directly received from respondents. The managers of the Basra oil firm served as the study's sample. The survey was given out to 167 people by the researchers. There were just 153 completed questionnaires received. 14 of them were disqualified as a result of incomplete questionnaires. 139 were used to create the final collection of data. The demographics of the participants are shown in Table 1. 0.11 percent of the sample's managers were high-level, followed by 0.30 percent of mid-level, and 0.59 percent of low-level managers. A master's degree or above was held by 0.24 percent of respondents, 0.76 percent received an undergraduate degree. The typical respondent lasted for roughly 8 years. Most participants were between the ages of 40 and 49.

Table 1. Demographics of the participations.

Demographics	Frequency (n = 139)	Percent %
Gender		
Male	99	.71
Female	40	.29
Job position		
High level	14	.11
Mid-level managers	40	.30
Low-level managers	85	.59
Age		
30–39	25	.20
40–49	58	.40
50–59	33	.22
≥60	23	.18
Academic qualifications		
Undergraduate	105	.76
Post-graduate	34	.24
Organizational tenure		
3–6	29	.23
7–10	46	.30
≥11	64	.47

The seven primary variables in the present study include perceived compatibility, perceived trialability, perceived complexity, perceived observability, and personal innovativeness as independent variable, user's satisfaction as mediative variable, while social sustainability of the metaverse is the dependent variable. This study adopts well-established scales from the existing literature to measure the variables in the research model. The current research operationalizes perceived factors using 13 items adapted from (Akour et al. 2022). as well as, used 3 items for user's satisfaction, adopted from (Simanjuntak and Purba 2020). In contrast, social sustainability of the metaverse was measured with 3 items adapted from (Arpaci et al. 2022). A 7-point scale (1 = strongly disagree, 7 = strongly agree) was utilized to operationalize all the concepts.

4   Data Analysis

According to Hair et al. (2010), the measurement model seeks to evaluate the validity and reliability of the created measures. To verify the convergent validity, the average extracted variance and composite reliability should be evaluated. Since they were greater than 0.50, as advised by Hair et al. (2010), the average variance extracted (AVE) values of all latent variables in this investigation ranged from 0.737 to 0.575, which was acceptable. Additionally, as shown by the results in Table 2, the outer loadings of all latent variables for all of the major constructs ranged between 0.918 and 0.737, above the acceptable value of 0.70 as advised by Hair et al. (2017). The latent variables' composite reliabilities also ranged between 0. 867 and 0. 726. These findings show that the measurement scales used into the model are highly reliable (Hair et al., 2010).

Table 2. Result of measurement model.

Variables	Items	LF	CR	AVE
Perceived traibility	PT1	0.807	0.849	0.737
	PT2	0.857		
	PT3	0.737		
Perceived observability	PO4	0.804	0.867	0.685
	PO5	0.869		
	PO6	0.808		
Users' compatibility	PC7	0.782	0.843	0.643
	PC8	0.766		
	PC9	0.801		
Perceived complexity	PX10	0.836	0.824	0.702
	PX11	0.881		
Personal innovativeness	PI12	0.750	0.826	0.613
	PI13	0.918		
User satisfaction	US14	0.739	0.813	0.592
	US15	0.824		
	US16	0.742		
Social sustainability of the Metaverse	SU17	0.814	0.726	0.575
	SU18	0.775		
	SU19	0.754		

As shown in Table 3 the Fornell and Larcker (1981) criterion was used to measure discriminant validity. The square root of the AVE is more than 0.5 of all constructs. This indicates that all the variables are distinctly different from the rest (Hair et al. 2010).

Table 3. Discriminant validity.

Variables	1	2	3	4	5	6	7
1. Perceived complexity	0.859						
2. Perceived observability	0.509	0.828					
3. Perceived traceability	0.506	0.655	0.802				
4. Personal innovativeness	0.556	0.553	0.568	0.838			
5. Social sustainability of the Metaverse	0.423	0.431	0.345	0.513	0.689		
6. User satisfaction	0.506	0.547	0.541	0.488	0.509	0.769	
7. Users' compatibility	0.519	0.652	0.561	0.463	0.411	0.456	0.783

This study has seven hypotheses to answer the study questions. To hypotheses, bootstrapping tool was used with 5000 bootstrap re-sampling (Hair et al. 2017). Table 4. Displays the outcomes of the structural model evaluation.

Table 4. Hypotheses test.

Direct Path	O	M	STDEV	O/STD	P
Perceived traibility - > User satisfaction	0.207	0.208	0.044	4.721	0.000
Perceived observability - > User satisfaction	0.216	0.215	0.049	4.453	0.000
Users' compatibility - > User satisfaction	0.037	0.038	0.049	0.758	0.224
Perceived complexity - > User satisfaction	0.206	0.207	0.044	4.706	0.000
Personal innovativeness - > User satisfaction	0.119	0.119	0.044	2.718	0.003
User satisfaction - > Metaverse	0.509	0.512	0.029	17.338	0.000
Indirect path					
Perceived traibility - > User satisfaction- > Metaverse	0.105	0.106	0.023	4.541	0.000
Perceived observability - > User satisfaction- > Metaverse	0.110	0.110	0.026	4.201	0.000
Users' compatibility - > User satisfaction - > Metaverse	0.019	0.020	0.025	0.749	0.227
Perceived complexity - > User satisfaction - > Metaverse	0.105	0.106	0.023	4.481	0.000
Personal innovativeness - > User satisfaction- > Metaverse	0.060	0.061	0.023	2.654	0.004

The results in Table 4 suggest a direct effect, but there was a strong indication that all statistical assumptions on direct effects were valid. Regarding the indirect influence hypotheses, the results revealed that the mediator variable plays a fully mediating role in the interaction between the dependent and the independent variables (Fig. 1).

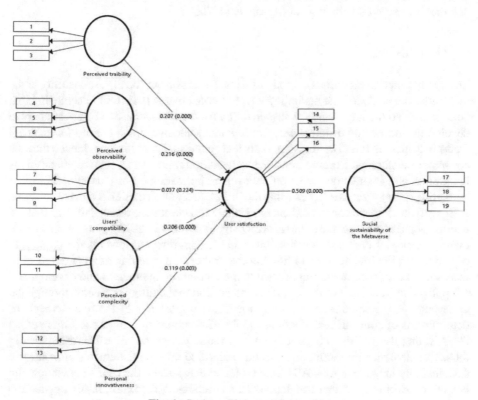

Fig. 1. Path coefficient of the model.

5 Discussion

The study's findings suggest that adoption properties features are crucial to the social sustainability of the metaverse. This lends weight to the hypotheses (H1–H5) that pre-supposed the consequences outlined above. This outcome is consistent with the theories advanced on the significance of the adoption properties elements impact. As a result, these findings supported (Akour et al. 2022), which provided a framework for under-standing the impact of adoption properties elements. After then, think about how to handle this situation given the ongoing advancement of technology. Therefore, businesses must seriously consider reducing user stress. Several actions need to be taken concerning technical enterprises; businesses should create a program to train all techni-cians, particularly those who work with users. On the other side, users need to be open to change and prepared for personal growth. This study's conclusion that user pleasure has an impact on the metaverse can be used to support the development of international trade. Regarding the seven hypotheses, the findings indicate that adoption properties elements and the Social Sustainability of the Metaverse have a mediating role for user pleasure. Can assert that user happiness contributes to the metaverse's continued viability. This outcome supported the findings of (Arpaci et al. 2022).

6 Conclusion

This study aimed to examine the impact of some factors on the social sustainability of the metaverse and analyzed the mediating effect of consumer satisfaction in achieving that goal. It adopted the importance of sustainability for the metaverse as a future strategy to develop the motivation of interaction in users and thus investigate the most significant factor in achieving this. The study suggests that social sustainability is achieved through consumer satisfaction. Because consumer satisfaction represents a state of unconscious integration and interaction based on metaverse applications. For example, there is a marketing strategy that is adopted using the SPICE promotion model. It aims to consume K-pop-oriented music content to lead the global market on metaverse platforms that are gaining popularity in the music services industry (Arpaci et al. 2022). Thus, it can be said that the choice of consumer satisfaction is an element directed towards the possibility of continuing the interaction and limiting the property of stopping at a certain level of reactions. It is of paramount importance to the continuity, interaction and development of applications that achieve the feature of integration in the offers presented through the metaverse. At the same time, the study found the importance of the five elements as determinants of consumer satisfaction, and therefore companies must exert more effort in activating the positive characteristics and reducing the negative characteristics in those five elements. Furthermore, this study identified important elements of metaverse sustainability in social terms. All of these elements represent the user's conviction and behavior resulting from that conviction. This is reflected, therefore, in his satisfaction with how to employ technical applications in a way that is in the interest of the user.

References

Abbas, S., et al.: Antecedents of trustworthiness of social commerce platforms: a case of rural communities using multi group SEM & MCDM methods. Electronic Commerce Research and Applications, 101322 (2023)

Agarwal, R., Prasad, J.: A conceptual and operational definition of personal innovativeness in the domain of information technology. Inf. Syst. Res. **9**(2), 204–215 (1998)

Akour, I.A., Al-Maroof, R.S., Alfaisal, R., Salloum, S.A.: A conceptual framework for determining metaverse adoption in higher institutions of gulf area: an empirical study using hybrid SEM-ANN approach. Comput. Educ. Artif. Intell. **3**(2022), 100052 (2022)

Al-Abrrow, H., Fayez, A.S., Abdullah, H., Khaw, K.W., Alnoor, A., Rexhepi, G.: Effect of open-mindedness and humble behavior on innovation: mediator role of learning. Int. J. Emerg. Markets (2021)

Alameri, M., Isaac, O., Bhaumik, A.: Factors influencing user satisfaction in UAE by using internet. Published Int. J. Emerging Technol. **10**, 8–15 (2019)

Albahri, A.S., et al.: Based on the multi-assessment model: towards a new context of combining the artificial neural network and structural equation modelling: a review. Chaos Solitons Fractals **153**, 111445 (2021)

Alfaisal Raghad, M., Zare, A., Alfaisal, A.M., Aljanada, R., Abukhalil, G.W.: The acceptance of metaverse system: a hybrid SEM-ML approach. Int. J. Adv. Appl. Comput. Intell. (IJAACI) **01**(01), 34–44 (2022)

AL-Fatlawey, M.H., Brias, A.K., Atiyah, A.G.: The role of Strategic Behavior in achievement the Organizational Excellence" Analytical research of the manager's views of Ur State Company at Thi-Qar Governorate". J. Administration Econ. **10**(37) (2021)

Al-Gahtani Said, S.: Computer technology adoption in Saudi Arabia: Correlates of perceived innovation attributes (2003)

Alnoor, A., et al.: How positive and negative electronic word of mouth (eWOM) affects customers' intention to use social commerce? a dual-stage multi group-SEM and ANN analysis. Int. J. Hum.–Comput. Interact., 1–30 (2022)

Alsalem, M.A., et al.: Rise of multiattribute decision-making in combating COVID-19: a systematic review of the state-of-the-art literature. Int. J. Intell. Syst. **37**(6), 3514–3624 (2022)

Ibrahim, A., Karatas, K., Kusci, I., Al-Emran, M.: Understanding the social sustainability of the Metaverse by integrating UTAUT2 and big five personality traits: a hybrid SEM-ANN approach. Technol. Soc. **71**(2022), 102–120 (2022)

Atiyah, A.G.: Impact of knowledge workers characteristics in promoting organizational creativity: an applied study in a sample of Smart organizations. PalArch's J. Archaeol. Egypt/Egyptology **17**(6), 16626–16637 (2020)

Atiyah, A.G.: The effect of the dimensions of strategic change on organizational performance level. PalArch's J. Archaeol. Egypt/Egyptol. **17**(8), 1269–1282 (2020)

Atiyah, A.G.: Effect of temporal and spatial myopia on managerial performance. J. La Bisecoman **3**(4), 140–150 (2022)

Atiyah, A.G.: Power Distance and Strategic Decision Implementation: Exploring the Moderative Influence of Organizational Context.

Atiyah, A.G.: Strategic Network and Psychological Contract Breach: The Mediating Effect of Role Ambiguity

Atiyah, A.G., Zaidan, R.A.: Barriers to using social commerce. In: Artificial Neural Networks and Structural Equation Modeling: Marketing and Consumer Research Applications, pp. 115–130. Springer, Singapore (2022)

Aubert, B.A., Hamel, G.: Adoption of smart cards in the medical sector: the Canadian experience. Soc Sci Med **53**(7), 879–894 (2001)

Bilgihan, A., Seo, S., Choi, J.: Identifying restaurant satisfiers and dissatisfiers: suggestions from online reviews. J. Hosp. Market. Manag.Manag. **27**(5), 601–625 (2018)

Anne, B., McDonagh-Philp., D.: Focus groups to support the industrial/product designer: a review based on current literature and designers' feedback **33**(1), 27–38 (2002)

Chew, X., Khaw, K.W., Alnoor, A., Ferasso, M., Al Halbusi, H., Muhsen, Y.R.: Circular economy of medical waste: novel intelligent medical waste management framework based on extension linear Diophantine fuzzy FDOSM and neural network approach. Environ. Sci. Pollution Res., 1–27 (2023)

Choi, Y., Zhang, L., Debbarma, J., Lee, H.: Sustainable management of online to offline delivery apps for consumers' reuse intention: focused on the meituan apps. Sustainability **13**(7), 3593 (2021)

Denis, J.L., Hebert, Y., Langley, A., Lozeau, D., Trottier, L.H.: Explaining diffusion patterns for complex health care innovations. Health Care Manage. Rev. **27**(3), 60–73 (2002)

Rogers, E.M.: Diffusion of innovations. Free Press, New York (2003)

Eneizan, B., Mohammed, A.G., Alnoor, A., Alabboodi, A.S., Enaizan, O.: Customer acceptance of mobile marketing in Jordan: an extended UTAUT2 model with trust and risk factors. Int. J. Eng. Bus. Manage. **11**, 1847979019889484 (2019)

Erdem, Seker, F.: Tourist experience and digital transformation. In: Research on Digital Communications, Internet of Things, and the Future of Cultural Tourism, pp. 103–120. IGI Global (2022)

Fadhil, S.S., Ismail, R., Alnoor, A.: The influence of soft skills on employability: a case study on technology industry sector in Malaysia. Interdiscip. J. Inf. Knowl. Manag.. J. Inf. Knowl. Manag. **16**, 255 (2021)

Gatea, A.A., Marina, V.: Higher education funding in Iraq in terms of the experience of particular developed countries. Int. J. Adv. Stud. **6**(1), 8–17 (2016)

Greenhalgh, T., Robert, G., Macfarlane, F., Bate, P., Kyriakidou, O.: Diffusion of innovations in service organizations: systematic review and recommendations. Milbank Q. **82**(4), 581–629 (2004)

Hair, J., Black, W., Babin, B., Anderson, R.: Multivariate Data Analysis: A Global Perspective. Multivariate Data Analysis: A Global Perspective, 7th edn. (2010)

Hamid, R.A., et al.: How smart is e-tourism? a systematic review of smart tourism recommendation system applying data management. Comput. Sci. Rev. **39**, 100337 (2021)

Hankookilbo (2022). https://m.hankookilbo.com/News/Read/A2022011413230003736

Hardgrave, B.C., Davis, F.D., Riemenschneider, C.K.: Investigating determinants of software developers' intentions to follow methodologies. J. Manag. Inf. Syst. **20**(1), 123–151 (2003)

Isaac, O., Abdullah, Z., Ramayah, T., Mutahar, A.M., Alrajawy, I.: Integrating user satisfaction and performance impact with technology acceptance model (TAM) to examine the internet usage within organizations in Yemen. Asian J. Inf. Technol. **17**(1), 60–78 (2018)

Jang, J., Yi, M.Y.: Modeling user satisfaction from the extraction of user experience elements in online product reviews. In: Proceedings of the CHI Conf, Extended Abstracts on Human Factors in Computing Systems, New York, NY, USA, pp. 1718–1725 (2017)

Jang, J.: Modeling user satisfaction from the extraction of user experience elements in online product reviews. In: Proc. of the 2017 CHI Conf, Extended Abstracts on Human Factors in Computing Systems, New York, NY, USA, pp. 1718–1725 (2017)

Johnson, V.L., Johnson, A.K., Washington, R., Torres, R.: Limitations to the rapid adoption of M-payment services: understanding the impact of privacy risk on M-Payment services. Comput. Hum. Behav.. Hum. Behav. (2018). https://doi.org/10.1016/j.chb2017.10.035

Asad, K., Masrek, M.N., Mahmood, K.: The relationship of personal innovativeness, quality of digital resources and generic usability with users' satisfaction. Digital Library Perspective © Emerald Publishing Limited 2059–5816 (2014). https://doi.org/10.1108/DLP-12-2017-0046

Khaw, K.W., et al.: Modelling and evaluating trust in mobile commerce: a hybrid three stage Fuzzy Delphi, structural equation modeling, and neural network approach. Int. J. Hum.-Comput. Interact. **38**(16), 1529–1545 (2022)

Lee L.H., Braud, T., Zhou, P., Wang, L., Xu, D., et al.: All one needs to know about metaverse: a complete survey on technological singularity, virtual ecosystem, and research agenda (2021)

Hong, L.S., Lee, H., Kim, J.H.: Enhancing the prediction of user satisfaction with metaverse service through machine learning. Comput. Mater. Continua Tech. Sci. Pres. (2022). https://doi.org/10.32604/cmc.2022.027943

Lee, Y.H.: Exploring key factors that affect consumers to adopt e-reading services. Unpublished Master thesis: Huafan University (2007)

Lee, Y.-H., Hsieh, Y.-C., Hsu, C.-N.: Adding innovation diffusion theory to the technology acceptance model: supporting employees' intentions to use e-learning systems. J. Educ. Technol. Soc. **14**(4), 124 (2011)

Yavuz, M., Corbacıoˇglu, E., Başoğlu, A.N., Daim, T.U., Shaygan, A.: Augmented reality technology adoption: case of a mobile application in Turkey. Technology in Society, vol. 66 (2021)

Manhal, M., Al-khalidi, A., Hamad, Z.: Strategic network: managerial myopia point of view. Manage. Sci. Lett. **13**(3), 211–218 (2023)

Martins, C.B.M.J., Steil, A.V., Todesco, J.L.: Factors influencing the adoption of the Internet as a teaching tool at foreign language schools. Comput. Educ.. Educ. **42**(4), 353–374 (2004)

Porter, M.E., Heppelmann, J.E.: Harvard business school publishing corporation (2015)

Rogers, E.M.: Diffusion of Innovations. Free Press, New York (2003)

Suzuki, S.N., Kanematsu, H., Barry, D.M., Ogawa, N., Yajima, K.: Virtual experiments in metaverse and their applications to collaborative projects: the framework and its significance. Procedia Comput. Sci. **176**, 2125–2132 (2020)

Shih, C.H.: Integrating Innovation Diffusion Theory and UTAUT to explore the influencing factors on teacher adopt e-learning system–with MOODLE as an example. Dayeh University. Unpublished Master thesis (2007)

Simanjuntak, D.C.Y., Purba, P.Y.: Peran mediasi customer satisfaction dalam customer experience dan loyalitas pelanggan. Jurnal Bisnis Dan Manajemen 7(2) (2020)

Sonnenwald, D.H., Maglaughlin, K.L., Whitton, M.C.: Using innovation diffusion theory to guide collaboration technology evaluation: Work in progress. In: Proceedings Tenth IEEE International Workshop on Enabling Technologies: Infrastructure for Collaborative Enterprises. WET ICE 2001, pp. 114–119 (2001)

Suzuki, S.N., Kanematsu, H., Barry, D.M., Ogawa, N., Yajima, K., et al.: Virtual experiments in metaverse and their applications to collaborative projects: The framework and its significance. Procedia Comput. Sci. **176**, 2125–2132

Tang, Y.: Help first-year college students to learn their library through an augmented reality game. J. Acad. Librariansh.Librariansh. **47**(1), 102294 (2021)

Tobbin, P.E.: Modeling adoption of mobile money transfer: A consumer behaviour analysis (2010)

Vess, L.J., Kiser, A., Washington, R., Torres, R.: Limitations to the rapid adoption of M-payment Services: understanding the impact of privacy risk on M-payment services, vol. 79, pp. 111–122, February 2018

Wang, Q., Khan, M.S., Khan, M.K.: Predicting user perceived satisfaction and reuse intentions toward massive open online courses (MOOCs) in the COVID-19 pandemic: An application of the UTAUT model and quality factors. Int. J. Res. Bus. Soc. Sci. (2147–4478) **10**(2), 1–11 (2021)

Wheelen, L., Hunger, D., Hoffman, N., Bamford E.: Strategic Management and Business Policy Globalization, Innovation and Sustainability. 15th Edition, Global Editions, Pearson Education Limited, New York, USA (2018)

Siqi, Y., Cai, X., Wang, Z., Zhang, Y., Luo, S., Jingdong, M.: Impact of gamification elements on user satisfaction in health and fitness applications: a comprehensive approach based on the Kano model. Computers in Human Behavior Volume 128, March 2022, 107106 (2022)

The Future of Metaverse in Improving the Quality of Higher Education: A Systematic Review

Marwa Al-Maatoq(✉) ⓘ, Munaf Abdulkadhim Mohammed ⓘ,
and Abdulridha Nasser Mohsin ⓘ

Technical College of Management, Sothern Technical University, Basra, Iraq
marwa.mousa993@stu.edu.iq

Abstract. The COVID-19 pandemic has had a significant negative impact on the educational system, resulting in an increased emphasis on expediting the process of digitalizing education. Therefore, the incorporation of Metaverse exhibits considerable promise in constructing a resilient framework for the future of higher education. The objective of this study is to examine the potential ramifications of Metaverse technology on higher education. More specifically, it aims to investigate the potential advantages that the higher education sector could derive from the envisioned virtual environment. The present study utilised a theoretical systematic review technique in accordance with the PRISMA approach to examine and synthesise extant material pertaining to the utilisation of the Metaverse in higher education. This study examines various concepts and terminology associated with the subject matter and explores their potential implications. The research findings presented in this study offer evidence to support the feasibility of incorporating virtual reality technology in the realm of higher education. This integration has the potential to better interactions between humans and computers, as well as improve the overall experience between students and teachers. To the best of the researcher's knowledge, this study is the first attempt to provide a thorough description of the educational Metaverse within the context of higher education. Additionally, the study aims to investigate the potential consequences of the educational Metaverse in the future. Furthermore, this research possesses practical significance as it has assisted educational authorities in understanding the importance of the Metaverse within the domain of higher education. As a result, this has enhanced their capacity to formulate objectives and strategies that are aligned with the concept of the Metaverse. Owing to the emerging nature of this particular field, there is a limited availability of relevant scholarly material. This work contributes to the existing body of knowledge and suggests the need for additional research.

Keywords: Metaverse · higher education · teaching platforms · avatar · virtual classroom · virtual reality · extended Reality

© The Author(s), under exclusive license to Springer Nature Switzerland AG 2023
M. Al-Emran et al. (Eds.): IMDC-IST 2024, LNNS 895, pp. 108–130, 2023.
https://doi.org/10.1007/978-3-031-51716-7_8

1 Introduction

The field of computer technology has witnessed notable advancements that exert a profound influence on our everyday lives by revolutionizing and enhancing social interactions, communication, and human touch (Mystakidis, 2022). The Metaverse is commonly considered the subsequent iteration of social interactions. The concept of a manufactured environment pertains to a constructed setting wherein individuals are able to reside in accordance with the regulations established by the creator (Hwang and Chien 2022). The concept of the Metaverse refers to a digitally enhanced domain that spans both the actual world and virtual space. The concept being discussed is the integration of tangible and concrete reality with digital technology, enabling individuals to envision numerous virtual representations of the physical world, whether they are actual or hypothetical, serving various objectives (Akour et al. 2022).

According to (Arpaci et al. 2022), the Metaverse can be conceptualized as a nascent virtual realm whereby individuals engage in social interactions through the use of personalized digital representations known as avatars. The Metaverse holds the potential to provide a multitude of opportunities across several sectors. To date, numerous studies (Hwang and Chien 2022) have acknowledged the Metaverse technology as an auspicious technological advancement in contemporary times. However, the incorporation of the Metaverse within the realm of education remains largely unexplored. Numerous instructors may exhibit a dearth of understanding regarding the defining attributes of the Metaverse, let alone the potential applications that might be derived from this burgeoning technological advancement. The education sector plays a vital role in both society and the economy of countries, particularly in those where traditional methods of instruction, such as content transfer, classroom-based learning, and reliance on textbooks, persist despite the presence of various technology advancements (Friesen 2017). In contemporary times, there exists a highly competitive pursuit to establish the fundamental framework, protocols, and norms that will serve as the governing principles for the Metaverse (Mystakidis 2022).

The year 2021 witnessed the emergence of the Metaverse, a conceptual framework denoting a digitally simulated reality wherein individuals can engage with a computer-generated environment and fellow users in immediate temporal synchrony. The emerging discipline of global Metaverse studies has resulted in the acknowledgement of the Metaverse as an up-and-coming educational phenomenon with considerable potential. The presence of the Metaverse is frequently linked to the advent of numerous innovative technologies (Zhang et al. 2022). However, the existing body of research has predominantly overlooked the examination of the Metaverse within the context of education, instead placing greater emphasis on the individual exploration of Metaverse-related technology in learning settings. There appears to be a paucity of understanding among educational scholars regarding the Metaverse, encompassing its various components and prospective uses within the realm of education. Therefore, the main aim of this academic paper is to conduct a critical analysis of a range of prominent scholarly works, with the purpose of offering a complete comprehension of the concept of the Metaverse within the realm of education. This examination encompasses several aspects, such as its definition, framework, characteristic attributes, prospective applications, obstacles, and areas for future investigation. This study presents a number of notable contributions, which can be succinctly stated as follows.

The primary purpose of this article is to analyze the Metaverse in the context of higher education methodically and comprehensively. The discussion begins with a deep dive into the background of the Metaverse, then moves on to a precise explanation of its conceptualization and an in-depth exploration of its defining traits. Following this, we do a thorough analysis of the most recent studies on these topics, providing a multifaceted look at the subject.

Second, the purpose of this study is to present a comprehensive theoretical ground-work for employing the Metaverse in the context of higher education. To round out this strategy, we will examine how Metaverse-based learning differs from both traditional classroom settings and screen-based online education. Third, this section will explore the potential uses, issues, and future study topics for the Metaverse in the realm of educa-tion. This article seeks to provide clarification on the primary uncertainties surrounding the Metaverse and its utilization within the educational sphere. Specifically, the subse-quent research inquiries will serve as the basis for further investigations pertaining to the implementation of Metaverses in educational contexts.

Q1: "How can the concept of Metaverse be delineated?

Q2: "What are the applications of the Metaverse in the field of education?

Q3: What are the merits and demerits associated with the utilization of Metaverse in the realm of education?

Q4: "What are the obstacles and opportunities that emerge from the deployment of the Metaverse in the field of education?

Q5: What is the rationale behind employing the Metaverse in the realm of education?

Q6: Does the Metaverse educational environment exhibit superiority over traditional education?

An in-depth familiarity with the existing and future developments of the Metaverse in the context of higher education is essential for an efficient exploration of this topic. This article presents a comprehensive literature analysis on the topics of Metaverse, online learning, and the roles of teachers and students in a virtual classroom. The current inves-tigation follows the following structure to accomplish its goal: In Sect. 2, we examine the relevant literature and present primary definitions of critical topics. In Sect. 3, we briefly explain the paper's questions. In Sect. 4, we get a simplified explanation of the PRISMA framework. Following this, in Sect. 5, we explain the techniques used to con-duct the systematic review. Specifically designated for the presentation and discussion of the results in Sect. 6 of the text. Section 7 presents the results and acts as the study's conclusion.

2 Literature Review

2.1 The History of Metaverse and the Virtual Environment

In the last month of 2021, Mark Zuckerberg unveiled Meta as the new corporate identity for Facebook, the leading social networking platform, with the aim of spearheading the realization of the Metaverse concept. According to (Koohang et al. 2023; Atiyah 2023), it is posited that Meta will serve as the forthcoming advancement in social connection, fundamentally altering the manner in which individuals establish and maintain relationships with their acquaintances and loved ones, engage in various activities, and foster the growth of entrepreneurial endeavors. As per Zuckerberg's perspective, the fundamental characteristic of the Metaverse concept is in the experience of presence and the capacity to engage with the virtual realm fully. The potential trajectory of Metaverse platforms has been conceptualized with the aim of creating three-dimensional representations of residences, workplaces, and tourist sites. According to (Azmi et al. 2023), the utilization of virtual simulations would enable users to engage with the environment remotely, hence allowing their presence even in the absence of physical proximity.

The concept of the Metaverse was initially introduced in a science-fiction book titled Snow Crash as a means to depict a fully immersive three-dimensional virtual environment (Arpaci and Bahari 2023). The phrase in question has undergone a developmental process spanning nearly three decades. The film Ready Player One, released in 2018, reinvigorated scholarly conversations surrounding the Metaverse notion. The video depicts a virtual realm called 'OASIS' wherein individuals can connect to this digital environment, assume personalized avatars, and engage in unrestricted activities within the framework of established guidelines. The film Ready Player One depicts various technologically advanced concepts that appear to be within reach, such as the utilization of head-mounted displays (HMDs) for virtual reality (VR) immersion, widespread sensing capabilities, haptic feedback systems, and the ability to model the physical world. These technological breakthroughs offer the general public realistic opportunities to actualize the notion of the Metaverse (Duan et al. 2021).

The phrase 'Metaverse' is formed by combining the words 'meta' and 'universe' and refers to a virtual reality environment that is three-dimensional and lays considerable importance on interpersonal interactions. Within this digital realm, individuals engage in communication either through avatars or as active participants (Koo et al. 2022). The Metaverse can be defined as a virtual domain that incorporates a diverse array of activities, surpassing the boundaries of social interaction and games. It serves as a platform for many endeavors, such as commerce, education, cultural exchange, employment, and societal development, enabling individuals to both participate in and contribute to this alternate universe. The Metaverse, although shown as an innovative technological product, essentially embodies a transforming phenomenon that elucidates the use of technology and its integration into the very essence of human existence. A further issue pertaining to the Metaverse revolves around potential disparities in time perception due to users' diminished bodily awareness during virtual reality experiences (Arpaci et al. 2022).

The Metaverse remains a dynamic notion that continues to undergo evolution, with various individuals contributing to its interpretation and development. From a technical standpoint, the manner in which humans engage in communication is consistently advancing. Consequently, there has been a notable emergence of technological innovation, characterized by the integration of multiple novel technologies and the advent of new applications for the Internet. Metaverse has demonstrated enormous commercial potential from a commercial standpoint. In recent times, prominent corporations have directed their efforts towards the establishment of the Metaverse as a novel avenue for capital exportation. From a broader viewpoint, the progression of a Metaverse's evolution can be categorized into three sequential stages: (i) digital twins, (ii) digital natives, and ultimately (iii) surreality (Wang et al. 2022).

To accurately portray the real world, the first step is to generate large-scale, high-fidelity digital twins of people and objects in virtual settings (Lv et al. 2022). In the second phase, native content creation takes center stage, with the help of avatars representing digital natives who may work independently in virtual environments to come up with new ideas and gain new perspectives on the digital world. In its final form, the Metaverse is an autonomous, surreal reality environment that consumes the real world. In this stage, the virtual world will surpass the real one, making it possible for more scenes and lives to exist online than in the offline world (Wang et al. 2022).

2.2   The Concept of Metaverse

Due to their impact on social behaviour, interpersonal communication, and group dynamics, internet and computer-based technologies have enormous practical importance. In our daily lives, the development of computer technology occurs in stages. First, we see the development of personal computers, and then we see the introduction of the Internet. The third and last stage is the development of mobile devices. Currently, we are in a period characterised by the widespread use of virtual reality technologies to create fully immersive worlds for a variety of purposes (Saritas and Topraklikoglu 2022; Atiyah and Zaidan 2022).

Presently, the global dissemination of technical advancements is evident, with the Metaverse serving as a prominent illustration. The term 'Metaverse' pertains to a theoretical manifestation of the internet that spans a cohesive, omnipresent, and immersive virtual domain. This is facilitated by the employment of virtual reality (VR) and augmented reality (AR) headsets. (Setiawan and Anthony 2022). A wide range of diverse notions has characterized the term the Metaverse, which refers to a computer-generated cosmos. These include lifelogging, the notion of a collective space within virtuality, the idea of an embodied internet or spatial internet, the concept of a mirror world, and the concept of an omniverse, which serves as a platform for simulation and cooperation (Lee et al. 2021). There are numerous noteworthy definitions pertaining to the present subject matter. Although we cannot endorse all the ideas presented in these sources, it is worth noting a few that provide insights into the definition of the 'Metaverse.'

The phrase 'Metaverse' is coined by combining the terms 'meta', which signifies a state of transcendence, and 'verso,' derived from the notion of the cosmos. The emergence of the digital world is a consequence of the integration of various technologies. The objective is to achieve complete integration between the digital realm and reality, enabling the transfer of all activities and parameters from the physical world to the virtual space (Contreras et al. 2022). When people talk about the 'Metaverse,' they are usually referring to a virtual world where they can take on various roles and participate in a variety of activities while hiding behind an online persona, or 'avatar' Park and Kim (2022) argue that XR acts as a bridge between the virtual world of avatars and the real world of their users. Lee et al. (2021) explain that the word 'Metaverse' refers to a virtual world where real and digital worlds collide. The development of the World Wide

Web (WWW) and Extended Reality (XR) has paved the way for this amalgamation of systems.

The term 'Metaverse' refers to an expansive virtual environment that contains a range of elements, including avatars that users, digital objects, virtual landscapes, and numerous computer-generated components may control. Within this domain, individuals (represented by their avatars) are able to utilize their virtual identities via smart devices in order to engage in communication, collaboration, and social interaction with fellow users. The integration of the Metaverse encompasses the tripartite realms of the physical, human, and digital domains. In the subsequent discourse, we expound upon the intricate interconnections among the three realms, the constituent elements inside the Meta-verse, and the intricate dynamics of information dissemination within the Metaverse, as elucidated by Wang et al. (2022).

3 Education and the Metaverse

3.1 Development of Modern Education Revolution

The field of education is of great importance to both society and the economy. Despite the advent of several technological breakthroughs, the fundamental methods employed in education, such as content transmission, classroom instruction, and the use of textbooks, have remained essentially constant (Friesen 2017a, b; Gatea and Marina 2016). The evolution of the educational system spans several centuries and has undergone continuous modifications to align with the prevailing methodologies. In their study (Lin et al. 2022; Alnoor et al. 2022a, b) conducted a straightforward comparison of three instructional approaches, as presented in Table 1.

It is widely acknowledged that the primary objective of Basic education in con-ventional schools is to equip students with fundamental literacy and numeracy skills, along with essential knowledge and competencies that align with their developmental stages. Additionally, this form of education aims to adequately prepare students for fur-ther academic pursuits at the tertiary level. The primary objective of basic education is to enhance students' knowledge and skills acquired inside educational institutions, hence facilitating their personal growth, societal integration, and citizenship develop-ment (Shaturaev 2021). With approximately 118,000 infections in 114 countries over the course of three months, the World Health Organisation (WHO) has designated coro-navirus (COVID-19) a pandemic (WHO, 2020b). Educators, students, and families can all benefit from using distance learning to reduce the spread of the coronavirus. The need has driven the development of a number of methods, including the use of online educational platforms, including Blackboard, Zoom, TronClass, Classin, and WeChat group platforms. In both industrialised and developing countries, the use of Internet platforms for educational instruction and acquisition is not a novel technique. Educa-tors, students, families, and the government all face substantial hurdles as they make the shift from traditional in-person classes to online learning. Financial constraints, a lack of suitable skills, an absence of an adequate information and communication technology infrastructure, a lack of access to the internet, and a dearth of supplementary educational materials are just some of the causes of these difficulties (Basilaia and Kvavadze 2020).

Distance e-learning is a practical strategy for keeping the educational system going, say Tadesse and Muluye (2020).

The e-learning sector has seen substantial transformation since the advent of the internet and the widespread use of computers in the late 1990s. According to industry analysts, it is currently contended that the third generation of computers has emerged, facilitated by the second wave of mobile computing and social media, which has brought about the concept of microlearning through the utilization of short video-based learning sessions. The Metaverse is a dynamic virtual environment that operates continuously, offering a three-dimensional space for social interaction with acquaintances, thereby supplanting the conventional, two-dimensional web pages and contact lists found on personal computers. The modification above carries significant ramifications in the context of acquiring the skills necessary for constructing features (Azoury and Hajj 2023).

Virtual education was originally utilized in the US as part of e-learning to hasten its implementation and allow students to study it online. Virtual education uses the Internet, audio, video, multimedia, electronic books, Email, Facebook, and WhatsApp chat groups to give the student the knowledge he needs in his chosen specialization to improve his scientific level.

However, AI in education has struggled to flourish since education systems worldwide are more resistant to technological advances in their traditional organization. AI was promised to personalize instruction through tutor systems. This promise is being realized as technology experiments with diverse models worldwide (Pedroet al. 2019).

Table 1. Comparison of three types of education.

Comparison Factors	Traditional Education	Online Education	Metaverse Education
Location	School	School, home	"School, home"
Equipment	Book, pen, blackboard	Computer, mobile phone, tablet	"Brian-computer interface, wearable devices"
Teaching form	One–to–many	One- to- many, one- to- one	"One- to- many, one- to- one"
Educator	Teacher	Knowledge sharer	Knowledge sharer
Educated	Student	Learner	Learner
Teaching contact	Social and natural science	Interest, social and natural science	Customization
Teaching purpose	Personal training	Personal training enriches lives	All- round education
Teaching support	none	Web 2.0	Web 3.0

3.2 Development of Modern Education Revolution

Institutions of higher education are currently engaged in the active development of novel tactics aimed at reevaluating their approach to fulfilling their goal. The scrutiny surrounding the value of higher education has intensified due to economic and political pressures (Alexander et al. 2019). This scrutiny has been further exacerbated by the worldwide spread of COVID-19, which has been classified as a pandemic by the World Health Organization (Oda Abunamous et al. 2022), posing a significant threat to humanity (World Health Organization 2019).

In light of evolving circumstances, institutions of higher education are currently engaged in a process of reassessing their approaches to effectively address the academic and social requirements of all students pursuing certificates or degrees (Muhsen et al. 2023; Al-Hchaimi et al. 2023; Chew et al. 2023). The transition towards a student-centred learning approach necessitates the active involvement of both faculty members and academic advisers, who are required to assume the roles of guides and facilitators. The emergence of new degree programs, accompanied by the increasing prevalence of interdisciplinary studies, suggests that educational institutions are actively striving to offer students opportunities that foster connections between different fields of study. Moreover, this trend reflects a reevaluation of how to effectively leverage current resources in order to enhance the educational process of learners (Alexander et al. 2019).

Previously, a highly competitive environment existed in the pursuit of establishing the necessary infrastructure, protocols, and standards that would serve as the governing framework for the Metaverse. In order to attract customers and establish themselves as the prime destination for the Metaverse, significant corporations are aggressively attempting to develop their own proprietary hardware and software ecosystems (Mystakidis 2022). Technological progress is mainly responsible for the Metaverse's proliferation in the classroom. Thus, using modern technology tools and resources is crucial for the successful application of the Metaverse in educational contexts. Therefore, a wide range of technologies can provide the backbone infrastructure for the educational Metaverse, providing in-depth support for both traditional and virtual elements (Zhang et al. 2022; Abbas et al. 2023; Bozanic et al. 2023).

According to Lee (2021), the term 'Metaverse' refers to a realm where virtual and physical realities intersect and mutually influence one another, facilitating various social, economic, and cultural endeavors aimed at generating value. The amalgamation of the physical realm with virtual reality is not merely a straightforward fusion but rather a complex interplay. Moreover, the term 'Metaverse' encompasses a realm where routine existence and commercial endeavors are seamlessly integrated. With the rapid integration of the Metaverse into contemporary society, specific Metaverse applications have already found utility within the realm of education. The integration of technology in instructional practices has been of paramount importance in the realm of education and human development. Scholars and policymakers have developed multiple conceptual frameworks and analyzed various approaches pertaining to the utilization of emerging technologies in educational contexts (Singh et al. 2022).

This technology has the potential to be utilized for educational purposes in geographically remote as well as physically inaccessible regions. For instance, virtual reality technology finds application in various domains such as military training, specifically in engineering training involving nuclear studies, pilot and astronaut training utilizing virtual cockpits, language learning, and medical education involving the use of simulated cadavers (Tas and Bolat 2022). Because of its potential to overcome the inherent limitations of web-based, two-dimensional e-learning tools in terms of realism and motivation, metaverse technology holds great promise for generating interest in distance education. An increasingly important factor in the development of students' learning is the application of information and communication technologies (ICTs) as facilitative instruments for educational processes based on evolving paradigms. The demarcation of technology has prompted shifts in educational policy that prioritise its incorporation and utilisation (Saritas and Topraklikoglu 2022).

In educational and learning settings, students can benefit from the increased immersion and participation that Metaverse offers. As you will see below, the Metaverse will have a significant impact on the educational world for a number of reasons (Fitria and Simbolon 2022):

1. The learning process will be enhanced in terms of enjoyment and engagement.
2. Practical lessons can be conducted with a higher degree of realism.
3. The process of acquiring knowledge and transitioning into a knowledge producer can be initiated at the earliest opportunity.
4. The availability and expansion of learning resources will increase.
5. In contemporary times, educational institutions have transcended their conventional definition as mere physical structures. The advent of the Metaverse is anticipated to have a transformative impact on the physical infrastructure of educational institutions everywhere.

4 Methods

4.1 Research Design

In this study, we examine the literature on Metaverse uses in classrooms in great detail. Systematic literature reviews aim to identify research questions, assess relevant studies, and synthesise findings (Singh and Thurman 2019). The research followed the guidelines laid out by the PRISMA 2020 checklist, which is a generally acknowledged methodological framework for conducting meta-analyses and systematic reviews. Researchers can make sure their worldwide literature review is well-organized and well-documented by using the PRISMA checklist (Saritas and Topraklikoglu 2022).

4.2 Research Design

The field of education in Metaverse involved an examination of databases. The relevant scholarly works were also consulted to establish a foundation for identifying the keywords. This study employed the keyword 'Metaverse' together with combinations of keywords such as 'virtual reality,' 'education,' and 'e-learning.'

In order to find relevant results, we looked for them in the high-impact databases Web of Science and Scopus. The study did not put any time constraints on the literature assessment, which is understandable given the growing popularity of the Metaverse environment in academic settings in recent years (Alnoor et al. 2022a, b). The publications were carefully evaluated and added to the systematic review after being retrieved from databases. Predetermined inclusion and exclusion criteria, as well as the quality and relevance of the studies, guided this selection procedure.

5 Result

5.1 Research Design

A metareview was conducted to examine the methodologies and reporting practices employed by scholars on the subject of Metaverse technology and education while conducting systematic quantitative literature evaluations. In order to fulfil the objectives of this study, an extensive literature search was conducted with the aim of identifying systematic review studies. To be more precise, the publications above were published exclusively in Metaverse technology and education journals up until the conclusion of August 2023.

The methods outlined in Fig. 1 and detailed below adhere to the guidelines set out by the Preferred Reporting Items for Systematic Reviews and Meta-Analyses (PRISMA). The review employed the PRISMA methodology, which consisted of four sequential steps: (1) inclusion by keyword search, (2) eligibility assessment, (3) screening process, and (4) identification of relevant studies (Pahlevan-Sharif et al. 2019).

5.2 Inclusion

Upon conducting an extensive examination of several preliminary publications pertaining to the integration of Metaverse in educational settings, it became apparent that a dearth of scholarly literature exists in this domain. This scarcity can be attributed to the novelty and emerging nature of the subject matter. There was a limited quantity of recent scholarly articles pertaining to the problem. Consequently, the investigation commenced by first discovering empirical evidence and subsequently narrowing down the search history of relevant publications and prior scholarly contributions. Multiple database searches and search engines were utilized, including prominent platforms such as Google Scholar, ScienceDirect (Elsevier et al.), Taylor & Francis Online, Wiley Online Library, IEEEXplore, and Springer.

The search was refined by utilizing multiple keywords, as depicted in Fig. 1. The utilized keywords encompassed the topics of 'Metaverse and education', 'Metaverse in higher education', 'Metaverse's role in education', and 'Metaverse tools in education'. Furthermore, the search queries 'e-learning,' 'virtual reality,' and 'Metaverse' were predominantly employed to explore the database. It is important to highlight those certain keywords and phrases such as 'evaluation, online, and remote learning' were deliberately omitted. The preliminary search yielded a total of 206 items.

5.3 Eligibility

After applying a filtering process to the articles and documents obtained from the initial search, the research studies that were considered for inclusion in the review satisfied various criteria. It is worth noting that 14 duplicate articles were identified and subsequently eliminated from the dataset. The establishment of inclusion and exclusion criteria serves the purpose of evaluating the quality and pertinence of research within the existing literature, as depicted in Table 2.

Table 2. Inclusion and Exclusion Criteria

Inclusion Criteria	Exclusion Criteria
1. Book chapters, academic articles, and books 2. Research using the keyword Metaverse 3. Research including the term education 4. Learning or teaching 5. Full-text research 6. Public works 7. English-language studies	1. Whitepapers, online presentations, abstracts, and news 2. Studies that have not been released in English 3. Works that are not full texts 4. Studies whose topics aren't 5. Sufficiently significant to be looked at

5.4 Screening

The screening stage consisted of two sub-stages, namely filtering and an additional layer of exclusion. During the process of filtering, a total of 113 articles were initially reviewed and subsequently narrowed down. This was done in preparation for the final exclusion step, where 54 articles were deleted based on legitimate explanations. In this context, a rational explanation is categorized as an exclusion criterion, as outlined in Table 3.

5.5 Identification

The 25 articles chosen for this review were gathered and organized for synthesis, and their respective descriptions are provided in Table 3. The last phase of the selection process entails making preparations for the discussion segment.

Table 3. Some of the selected articles and studies with descriptions

No	Title	Refs	Aim of the Study
1	"Analyzing education based on Metaverse technology."	(Mustafa 2022)	The primary purpose of this research is to investigate the viability of Metaverse-based education by analysing current perceptions and identifying unmet educational needs
2	"Bibliometric Mapping of Metaverse in Education"	(Tas and Bolat 2022)	The purpose of this research is to provide a bibliometric visualisation of academic papers discussing educational applications of metaverse technology
3	"Students' Opinions about the Educational Use of the Metaverse"	(Talan and Kalinkara 2022)	The purpose of this research is to learn how students feel about using the Metaverse in the classroom
4	"Systematic Literature Review on the Use of Metaverse in Education"	(Saritas and Topraklikoglu 2022)	In this research, we look at how teachers have used Metaverse software. This research also delves into how the concept of the Metaverse has changed classroom dynamics over time
5	"Possibility of Metaverse in Education: Opportunity and Threat"	Fitria and Simbolon 2022).	The research team behind this project hopes to learn more about the advantages and disadvantages of using a Metaverse in the classroom
6	"Empowering Future Education: Learning in the Edu-Meraverse"	(Zhong and Zheng 2022)	The primary purpose of this research is to examine the Edu-metaverse framework and its potential educational applications. Next, this article investigates how public perception, technological limitations, and concerns about safety can stunt the development of Edu-metaverse programmes. These results have implications for the study of the Edu-Edu-metaverse…
7	"Metaverse in Education: Vision, Opportunities, and Challenges"	(Lin et al. 2022)	This research project investigates the current state of the technology, obstacles, opportunities, and future directions of using Metaverses in the classroom. This article aims to summarise the educational use of the Metaverse and investigate its inspirations
8	"Opportunities and challenges of metaverse for education: a literature review"	(Stanoevska-Slabeva 2022)	The research team behind this project hopes to learn more about the advantages and disadvantages of using a Metaverse in the classroom

(*continued*)

Table 3. (*continued*)

No	Title	Refs	Aim of the Study
9	"A conceptual framework for determining metaverse adoption in higher institutions of gulf area: An empirical study using hybrid SEM-ANN approach"	(Akour et al. 2022)	The purpose of this research is to probe Gulf area students' ideas on how to put the metaverse best to use in the classroom
10	"Students' Opinions about the Educational Use of the Metaverse"	(Talan and Kalinkara 2022)	The primary purpose of this research is to learn how students feel about using the Metaverse in the classroom
11	"The Importance of the Application of the Metaverse in Education"	(Contreras et al. 2022)	The primary goal of this essay is to present a thorough analysis of the metaverse and its potential applications in the realm of education
12	"Strategies for Implementing Metaverse in Education"	(Soni and Kaur 2023)	Metaverse education is the primary emphasis of this paper. The article contrasts the conventional classroom setting with the virtual one. The report also looks at how classrooms and learning tools have evolved through time
13	"Empirical Studies on the Metaverse-Based Education: A Systematic Review"	(Asiksoy 2023)	The goal of this research is to remedy the dearth of academic inquiry into the topic of education in Metaverse contexts by conducting a systematic review of the available empirical research on the subject
14	"Demystifying and Analysing Metaverse Towards Education 4.0"	(Raj et al. 2023)	The study's overarching goal is to contribute new knowledge and insight to the field of educational ecology through research
15	"Metaverse and education: Which pedagogical perspectives for the school of tomorrow"?	(Soriani and Bonafede 2023)	The study aims to provoke and respond to pressing questions and concerns about the metaverse's potential for use in formal and informal classroom settings. To what extent does the metaverse bring challenges and opportunities to the field of education?
16	"Use of Metaverse Technology in Education Domain"	(Rahman et al. 2023)	The objective of this study is to delineate the essential ideas and methodologies necessary for the transformation of the education sector through the utilization of the Metaverse

(*continued*)

Table 3. (*continued*)

No	Title	Refs	Aim of the Study
17	"Metaverse for education – Virtual or real"?	(Hussain 2023)	This article analyzes the current situation of the Metaverse in education and the anticipated future changes emerging from its integration. Finally, the author discusses the viability and challenges of integrating the Metaverse into schooling. I apologies, but I cannot respond without language or context. Provide more
18	"How Does the Metaverse Shape Education? A Systematic Literature Review"	(De Felice et al. 2023)	This study aims to fill this gap by offering a thorough analysis of the metaverse's influence on learning and teaching in the context of the labour market of the future
19	"Metaverse for Education: Developments, Challenges and Future Direction"	(Chamola et al. 2023)	The purpose of this research is to examine existing literature concerning Metaverse structure, categories, and constituents. After that, we will go into detail on how to put the Metaverse into action
20	"Metaverse for Education: Technical Framework and Design Criteria"	(Chen et al. 2023a, b)	The major goal of this essay is to provide a practical framework for developing and envisioning applications within the Education Metaverse. This effort hopes to aid in the development of Education Metaverse-wide uniformity
21	"A systematic literature review of the acceptability of the use of Metaverse in education over 16 years"	(Chua and Yu 2023)	The purpose of the PRISMA research is to examine trends in the use of Metaverse technology in educational settings from 2008 to 2022, assessing its perceived efficacy and ease of implementation
22	"Metaverse in Education: A Systematic Review"	(López Belmonte et al. 2023)	The purpose of this study is to synthesise the existing literature on the use of metaverses in education
23	Metaverse in Education: Contributors, Cooperation, and Research Themes	(Chen et al. 2023a, b)	The fundamental goal of this study is to use bibliometric analysis, social network analysis, topic modelling, and keyword analysis to identify the elements that contribute to scientific collaboration and research themes

(*continued*)

Table 3. (*continued*)

No	Title	Refs	Aim of the Study
24	Metaverse in education: A systematic literature review	(Pradana and Elisa 2023)	This article employs a systematic literature review strategy to present an extensive summary of prior research on the use of the metaverse in the field of education
25	The rising trend of Metaverse in education: challenges, opportunities, and ethical considerations	(Kaddoura and Al Husseiny, 2023)	The purpose of this essay is to survey the main issues, ethical worries, and possible hazards connected to teaching with the Metaverse. In addition, it hopes to suggest a structure for future studies that will investigate how the Metaverse may improve educational settings

6 Data Analysis and Discussion

6.1 Data Analysis

This study aims to identify scholarly articles that encompass the term Metaverse in conjunction with either education, e-learning, or virtual reality. Furthermore, the articles sought after are those published in reputable academic databases such as Scopus and Web of Science. The review omitted six research for which full texts were inaccessible and eight studies that were not written in the English language. Following a process of conformity assessment and quality evaluation conducted on the remaining 79 research, it was determined that 54 studies did not meet the criteria for inclusion in the review. Consequently, a total of 25 studies were deemed suitable and were included in the systematic review. Figure 2 depicts the flowchart illustrating the process of selecting papers that were incorporated in the review.

6.2 Distribution of Studies by Country

Regarding the distribution of studies across different countries, the research findings indicate that China had the most significant number of studies, totaling 44 (n = 44), as illustrated in Fig. 2. South Korea conducted the highest number of studies (n = 28), followed by India (n = 17), Indonesia (n = 11), Japan and Turkey (n = 10 each), Thailand (n = 8), and both Australia and Spain (n = 7 each).

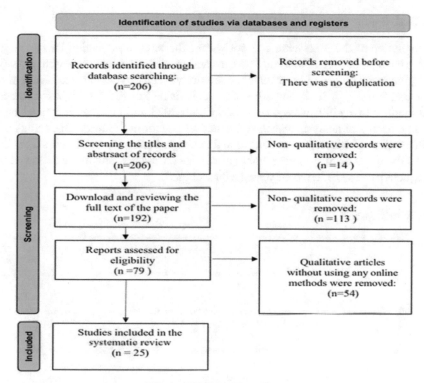

Fig. 1. PRISMA Flow Chart

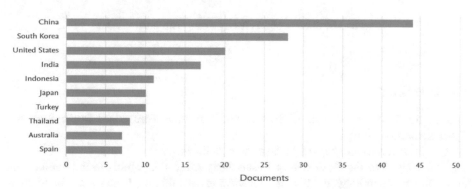

Fig. 2. Distribution of Studies by Country

6.3 Distribution of Studies by Country

The evolution of the Metaverse concept within the field of education has undergone notable changes between 2010 and 2023, as evidenced by scholarly research. However, a comprehensive review of the available literature yielded no research conducted during the years 2016, 2017, 2018, and 2019. The analysis reveals that the highest number of investigations (n = 199) were conducted in both 2023 and 2022, as indicated by the available data. The analysis reveals a fluctuating pattern in the research conducted on the integration of Metaverse in educational settings. Following an initial surge, there was a subsequent decline in the number of studies conducted, only to be followed by a subsequent resurgence in recent years (refer to Fig. 3).

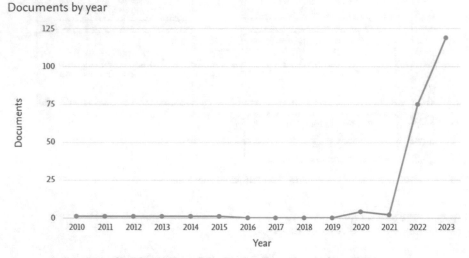

Fig. 3. Distribution of Studies by Years from 2010 to 2023

6.4 Discussion

This section provides answers to the five questions that were raised in the Research Question section above:

Q1: How can the concept of Metaverse be delineated?

The analysis of the aforementioned scholarly articles pertaining to the intersection of Metaverse technology and higher education reveals that the Metaverse can be characterised as an innovative technological framework that amalgamates physical reality and virtual environments through the utilization of three-dimensional (3D) realms and virtual reality. This integration enables individuals to assume virtual identities, known as avatars, within diverse virtual worlds (Albahri et al. 2021; Zainab et al. 2022).

Q 2: What are the applications of the Metaverse in the field of education?

Through the establishment of a Metaverse, an educational institution can construct a virtual rendition of its physical campus or school. Featuring various facilities such as classrooms, eating rooms, and instructors' rooms… Through the utilization of this method, individuals such as students, teachers, and other personnel associated with the institution are able to engage in efficient communication and interaction, simulating an in-person experience. This is achieved by means of video calls or videoconferences. This solution is particularly suitable for educational institutions that operate through e-learning platforms and provide comprehensive online training. It enables these institutions to enhance the human aspect of the educational experience despite the potential contradiction involved (Contreras et al. 2022; Albahri et al. 2022).

Q 3: What are the merits and demerits associated with the utilization of Metaverse in the realm of education?

According to Talan and Kalinkara (2022), the utilization of the Metaverse in educational settings is anticipated to become increasingly prevalent due to its pedagogical benefits. By implementing a range of strategies, an educational institution can establish a flexible environment that allows instructors, staff, and students to engage in effective communication (Atiyah 2023). This platform transcends the limitations typically associated with traditional classroom settings. In the digital realm, students can engage in communication with their lecturers effortlessly through the simple act of clicking a button. The Metaverse serves the purpose of encompassing a physical university or institution by transforming it into a virtual realm where educators, learners, and instructional models can engage in hybrid and collaborative educational sessions (Akour et al. 2022). Regarding the drawbacks, the utilization of Metaverse in education presents challenges in terms of learning efficacy and focus, leading to a disconnection between students and real-world experiences. Additionally, it has been seen to undermine classroom discipline (Talan and Kalinkara 2022).

Q4: What are the obstacles and opportunities that emerge from the deployment of the Metaverse in the field of education?

The utilization of Metaverse in the field of education presents numerous obstacles. The cost associated with computers and their components, including VR headsets, the requirement for 5G internet, the need for ample storage capacity for data and applications, and other technological advancements that facilitate individuals' engagement with the diverse virtual realms of the Metaverse, is regarded as a challenge that could impede the widespread adoption of this concept (Kaddoura & Al Husseiny 2023).

The potential application of Metaverse in higher education lies in its ability to provide immersive learning experiences through virtual reality. This technology facilitates interactive and experiential learning, enabling teachers and students to engage in practical and hands-on educational activities. The findings indicate a significant improvement in the rate of learning retention among pupils. Furthermore, it is noteworthy that educators possess the capability to gather distinct educational and evaluative data in order to gauge the efficacy of their instructional practices (Lin et al. 2022; Alsalem et al. 2022).

Q5: What is the rationale behind employing the Metaverse in the realm of education?

The Metaverse can be regarded as an indispensable technological tool for educational purposes due to its capacity as an instructional medium that has the potential to enhance cognitive flexibility in terms of creativity. This is achieved by enabling learners to engage in interactive exchanges with both peers and instructors. The utilization of avatars within the Metaverse realm offers a heightened level of anonymity in contrast to the offline domain. Consequently, this facilitates the expression of individual thoughts during problem-solving endeavors, thereby circumventing the influence of preconceived notions or external perspectives. The study conducted by Mustafa (2022) aims to enhance learners' creativity through the utilization of the Metaverse as an educational platform.

Q6: Does the Metaverse educational environment exhibit superiority over traditional education?

In the realm of conventional education, educators frequently employ various strategies to establish contextual frameworks, such as utilizing visual aids, narrating anecdotes, and presenting audiovisual materials. Following the advent of the Metaverse and advancements in virtual reality technology, it remains challenging to achieve a seamless merger of virtual and physical realities despite notable enhancements in the creative process. The educational Metaverse utilizes virtual reality (VR) technology to offer learners wearable devices that enable a comprehensive sensory experience. This immersive learning environment facilitates natural and contextually relevant learning experiences for individuals. Virtual reality (VR) technology has proven to be highly advantageous in various academic disciplines, including history, geography, biology, and medicine. By simulating real-life surroundings, VR enables learners to transcend traditional classroom settings and engage with subject matter in a more immersive manner. Moreover, VR facilitates the safe and unrestricted exploration of experimental equipment and observation of experimental phenomena within virtual laboratory settings (Zhong and Zheng 2022).

7 Conclusion

The advent of the Metaverse, spearheaded by prominent technology corporations, necessitates the readiness of the education sector to embrace this emerging technological paradigm. The implementation of the Metaverse is anticipated to yield beneficial outcomes within the realm of education. This strategy facilitates the efficient utilization of time by teachers and students for educational purposes. This study aims to examine two significant yet often neglected factors: the metaverse of higher education and the quality of higher education. These elements represent a relatively novel research domain within the subject of business studies.

Based on a thorough analysis of prior research conducted on Metaverse in the field of education, it can be inferred that Metaverse holds significant potential for enhancing the quality of implementation in higher education. The utilization of audiovisual-based educational technology is a prevalent and far-reaching implementation of the Metaverse concept. This application holds significant implications for the realm of learning. In the context of practicum-based education, the mere act of reading and observing is insufficient; active engagement and experiential involvement are necessary to comprehend and internalize the subject matter truly. The efficacy of learning is enhanced when it is integrated with the learning process itself, as well as through direct experience or simulation. The utilization of contemporary technologies and instruments has facilitated the enhancement of teacher efficacy and productivity, leading to an elevated standard of instructional excellence. The metaverse has enhanced students' learning experiences by facilitating the customization and personalization of educational resources to align with individual students' needs and skills. The metaverse has significantly influenced the quality of education, namely in the domains of administration, instruction, and learning inside educational institutions.

The most critical future recommendations of this study are as follows: First, the necessity of adopting metaverse technology in higher education for its effective role in achieving the best study environment for the student and teacher. Second: The need for higher education institutions to adopt metaverse technology in order to enhance the quality of university education.

References

Abbas, S., et al.: Antecedents of trustworthiness of social commerce platforms: a case of rural communities using multi group SEM & MCDM methods. Electronic Commerce Research and Applications, 101322 (2023)

Akour, I.A., Al-Maroof, R.S., Alfaisal, R., Salloum, S.A.: A conceptual framework for determining Metaverse adoption in higher institutions of gulf area: an empirical study using hybrid SEM-ANN approach. Comput. Educ. Artif. Intell. **3**, 100052 (2022)

Albahri, A.S., et al.: Integration of fuzzy-weighted zero-inconsistency and fuzzy decision by opinion score methods under a q-rung orthopair environment: a distribution case study of COVID-19 vaccine doses. Comput. Stand. Interf. **80**, 103572 (2022)

Albahri, A.S., et al.: Based on the multi-assessment model: towards a new context of combining the artificial neural network and structural equation modelling: a review. Chaos Solitons Fractals **153**, 111445 (2021)

Alexander, B., et al.: Horizon report 2019 Higher Education edition, pp. 3–41. EDU19 (2019)

Al Hchaimi, A.A.J., Sulaiman, N.B., Mustafa, M.A.B., Mohtar, M.N.B., Hassan, S.L.B.M., Muhsen, Y.R.: A comprehensive evaluation approach for efficient countermeasure techniques against timing side-channel attack on MPSoC-based IoT using multi-criteria decision-making methods. Egyptian Inform. J. **24**(2), 351–364 (2023)

Alnoor, A., et al.: Uncovering the antecedents of trust in social commerce: an application of the non-linear artificial neural network approach. Competitiveness Rev. Int. Bus. J. **32**(3), 492–523 (2022)

Alnoor, A., et al.: How positive and negative electronic word of mouth (eWOM) affects customers' intention to use social commerce? A dual-stage multi group-SEM and ANN analysis. Int. J. Hum.–Comput. Interac., 1–30 (2022)

Alsalem, M.A., et al.: Rise of multiattribute decision-making in combating COVID-19: a systematic review of the state-of-the-art literature. Int. J. Intell. Syst.Intell. Syst. **37**(6), 3514–3624 (2022)

Arpaci, I., Bahari, M.: Investigating the role of psychological needs in predicting the educational sustainability of Metaverse using a deep learning-based hybrid SEM-ANN technique. Interactive Learning Environments, 1–13 (2023)

Asiksoy, G.: Empirical studies on the metaverse-based education: a systematic review. Int. J. Eng. Pedagogy **13**(3) (2023)

Atiyah, A.G.: Power Distance and Strategic Decision Implementation: Exploring the Moderative Influence of Organizational Context.

Atiyah, A.G.: Strategic Network and Psychological Contract Breach: The Mediating Effect of Role Ambiguity

Atiyah, A.G., Zaidan, R.A.: Barriers to using social commerce. In: Artificial Neural Networks and Structural Equation Modeling: Marketing and Consumer Research Applications, pp. 115–130. Springer Nature Singapore (2022)

Azmi, A., Ibrahim, R., Ghafar, M.A., Rashidi, A.: Metaverse for real estate marketing: the impact of virtual reality on satisfaction, perceived enjoyment and purchase intention (2023)

Azoury, N., Hajj, C.: Perspective Chapter: The Metaverse for Education (2023)

Basilaia, G., Kvavadze, D.: Transition to online education in schools during a SARS-CoV-2 coronavirus (COVID-19) pandemic in Georgia. Pedagogical Res. **5**(4) (2020)

Bozanic, D., Tešić, D., Puška, A., Štilić, A., Muhsen, Y.R.: Ranking challenges, risks and threats using Fuzzy Inference System. Decision Making: Applications in Management and Engineering **6**(2), 933–947 (2023)

Buchholz, F., Oppermann, L., Prinz, W.: There is more than one Metaverse. i-com **21**(3), 313–324 (2022)

Chamola, V., et al.: Metaverse for Education: Developments, Challenges and Future Direction (2023)

Chen, X., Zhong, Z., Wu, D.: Metaverse for education: technical framework and design criteria. IEEE Trans. Learn. Technol. (2023)

Chen, X., Zou, D., Xie, H., Wang, F.L.: Metaverse in education: contributors, cooperations, and research themes. IEEE Trans. Learn. Technol. (2023)

Chew, X., Khaw, K.W., Alnoor, A., Ferasso, M., Al Halbusi, H., Muhsen, Y.R.: Circular economy of medical waste: novel intelligent medical waste management framework based on extension linear Diophantine fuzzy FDOSM and neural network approach. Environmental Science and Pollution Research, 1–27 (2023)

Chua, H.W., Yu, Z.: A systematic literature review of the acceptability of the use of Metaverse in education over 16 years. J. Comput. Educ., 1–51 (2023)

Contreras, G.S., González, A.H., Fernández, M.I.S., Martínez, C.B., Cepa, J., Escobar, Z.: The importance of the application of the Metaverse in education. Mod. Appl. Sci. **16**(3), 1–34 (2022)

De Felice, F., Petrillo, A., Iovine, G., Salzano, C., Baffo, I.: How does the metaverse shape education? a systematic literature review. Appl. Sci. **13**(9), 5682 (2023)

Duan, H., Li, J., Fan, S., Lin, Z., Wu, X., Cai, W.: Metaverse for social good: a university campus prototype. In: Proceedings of the 29th ACM International Conference on Multimedia, pp. 153–161, October 2021

Fitria, T.N., Simbolon, N.E.: Possibility of Metaverse in education: opportunity and threat. SOSMANIORA: Jurnal Ilmu Sosial dan Humaniora **1**(3), 365–375 (2022)

Friesen, N.: The textbook and the lecture: Education in the age of new media. JHU Press (2017)

Gatea, A.A., Marina, V.: Higher education funding in Iraq in terms of the experience of particular developed countries. Int. J. Adv. Stud. **6**(1), 8–17 (2016)

Hussain, S.: Metaverse for education–Virtual or real? In: Frontiers in Education, vol. 8, p. 1177429. Frontiers, April 2023

Hwang, G.J., Chien, S.Y.: Definition, roles, and potential research issues of the Metaverse in education: an artificial intelligence perspective. Comput. Educ. Artif. Intell. **3**, 100082 (2022)

Koo, C., Kwon, J., Chung, N., Kim, J.: Metaverse tourism: conceptual framework and research propositions. Current Issues in Tourism, pp. 1–7 (2022)

Koohang, A., et al.: Shaping the Metaverse into reality: a holistic multidisciplinary understanding of opportunities, challenges, and avenues for future investigation. J. Comput. Inf. Syst.Comput. Inf. Syst. **63**(3), 735–765 (2023)

Lee, L.H., et al.: All one needs to know about Metaverse: A complete survey on technological singularity, virtual ecosystem, and research agenda. arXiv preprint arXiv:2110.05352 (2021)

Lee, S.: Log in Metaverse: the revolution of human × space × time (IS-115). Seongnam: Software Policy & Research Institute (2021)

Lin, H., Wan, S., Gan, W., Chen, J., Chao, H.C.: Metaverse in education: Vision, opportunities, and challenges. In: 2022 IEEE International Conference on Big Data (Big Data), pp. 2857–2866. IEEE, December 2022

López Belmonte, J., Pozo-Sánchez, S., Moreno-Guerrero, A.J., Lampropoulos, G.: Metaverse in education: A systematic review (2023)

Lv, Z., Shang, W.L., Guizani, M.: Impact of digital twins and metaverse on cities: history, current situation, and application perspectives. Appl. Sci. **12**(24), 12820 (2022)

Muhsen, Y.R., Husin, N.A., Zolkepli, M.B., Manshor, N., Al-Hchaimi, A.A.J.: Evaluation of the routing algorithms for NoC-based MPSoC: a fuzzy multi-criteria decision-making approach. IEEE Access (2023)

Mustafa, B.: Analysing education based on Metaverse technology. Technium Soc. Sci. J. **32**, 278 (2022)

Mystakidis, S.: Metaverse. Encyclopedia **2**(1), 486–497 (2022)

Oda Abunamous, M., Boudouaia, A., Jebril, M., Diafi, S., Zreik, M.: The decay of traditional education: a case study under covid-19. Cogent Educ. **9**(1), 2082116 (2022)

Pahlevan-Sharif, S., Mura, P., Wijesinghe, S.N.: A systematic review of systematic reviews in tourism. J. Hosp. Tour. Manag.Manag. **39**, 158–165 (2019)

Park, S.M., Kim, Y.G.: A Metaverse: taxonomy, components, applications, and open challenges. IEEE Access **10**, 4209–4251 (2022)

Pedro, F., Subosa, M., Rivas, A., Valverde, P.: Artificial intelligence in education: Challenges and opportunities for sustainable development (2019)

Pradana, M., Elisa, H.P.: Metaverse in education: a systematic literature review. Cogent Soc. Sci. **9**(2), 2252656 (2023)

Rahman, K.R., Shitol, S. K., Islam, M.S., Iftekhar, K.T., Pranto, S.A.H.A.: Use of metaverse technology in the education domain. J. Metaverse **3**(1), 79–86 (2023)

Raj, A., Sharma, V., Rani, S., Singh, T., Shanu, A.K., Alkhayyat, A.: Demystifying and analysing metaverse towards education 4.0. In: 2023 3rd International Conference on Innovative Practices in Technology and Management (ICIPTM), pp. 1–6. IEEE, February 2023

Saritas, M.T., Topraklikoglu, K.: Systematic literature review on the use of metaverse in education. Int. J. Technol. Educ. **5**(4), 586–607 (2022)

Setiawan, K.D., Anthony, A.: The essential factor of Metaverse for business based on 7 layers of Metaverse–systematic literature review. In: 2022 International Conference on Information Management and Technology (ICIMTech), pp. 687–692. IEEE, August 2022

Shaturaev, J.: 2045: Path to nation's golden age (Indonesia Policies and Management of Education). Sci. Educ. **2**(12), 866–875 (2021)

Singh, J., Malhotra, M., Sharma, N.: Metaverse in education: An overview. Applying metalytics to measure customer experience in the Metaverse, 135–142. (2022)

Singh, V., Thurman, A.: How many ways can we define online learning? a systematic literature review of definitions of online learning (1988–2018). Am. J. Distance Educ. **33**(4), 289–306 (2019)

Soni, L., Kaur, A.: Strategies for implementing metaverse in education. In: 2023 International Conference on Disruptive Technologies (ICDT), pp. 390–394. IEEE, May 2023

Soriani, A., Bonafede, P.: Metaverse and education: which pedagogical perspectives for the school of tomorrow? J. Inclusive Methodol. Technol. Learn. Teach. **3**(1) (2023)

Stanoevska-Slabeva, K.: Opportunities and challenges of metaverse for education: a literature review. EDULEARN22 Proceedings, 10401–10410 (2022)

Tadesse, S., Muluye, W.: The impact of COVID-19 pandemic on education system in developing countries: a review. Open J. Soc. Sci. **8**(10), 159–170 (2020)

Talan, T., Kalinkara, Y.: Students' opinions about the educational use of the metaverse. Int. J. Technol. Educ. Sci. **6**(2), 333–346 (2022)

Tas, N., Bolat, Y.İ: Bibliometric mapping of Metaverse in education. Int. J. Technol. Educ. **5**(3), 440–458 (2022)

Wang, Y., et al.: A survey on metaverse: fundamentals, security, and privacy. IEEE Commun. Surv. Tutorials. WHO (2020b). Archived: WHO Timeline—COVID-19. World Health Organization

Yoo, K., Welden, R., Hewett, K., Haenlein, M.: The merchants of meta: A research agenda to understand the future of retailing in the Metaverse. J. Retailing (2023)

Zainab, H.E., Bawany, N.Z., Imran, J., Rehman, W.: Virtual dimension—a primer to Metaverse. IT Professional **24**(6), 27–33 (2022)

Zhang, X., Chen, Y., Hu, L., Wang, Y.: The Metaverse in education: definition, framework, features, potential applications, challenges, and future research topics. Front. Psychol. **13**, 6063 (2022)

Zhong, J., Zheng, Y.: Empowering future education: Learning in the Edu-Metaverse. In: 2022 International Symposium on Educational Technology (ISET), pp. 292–295. IEEE, July 2022

Integrating Ideal Characteristics of Chat-GPT Mechanisms into the Metaverse: Knowledge, Transparency, and Ethics

Abbas Gatea Atiyah[1], NimetAllah Nasser Faris[2], Gadaf Rexhepi[3(✉)], and Alaa Jabbar Qasim[4]

[1] Faculty of Administration and Economic, University of Thi-Qar, Nasiriyah, Iraq
abbas-al-khalidi@utq.edu.iq
[2] College of Arts, Department of English, University of Basrah, Basrah, Iraq
nimetullah.faris@uobasrah.edu.iq
[3] Southeast European University, Tetovo, North Macedonia
g.rexhepi@seeu.edu.mk
[4] School of Computing, College of Arts and Sciences, Universiti Utara Malaysia, 06010 Sintok, Kedah, Malaysia
alaa_jabbar@ahsgs.uum.edu.my

Abstract. Chat-GPT is a very advanced language model in artificial intelligence. It can be used in administrative and economic research and has many other applications. On the other hand, the metaverse is a wide and expansive space that expresses a virtual reality about the nature of the service or product that was recently produced or developed. It is possible to integrate these two tools by introducing the ideal mechanisms of Chat-GPT into metaverse applications. This study deals with a logical presentation to form a research framework on the process of integrating Chat-GPT mechanisms into metaverse applications. By referring to the literature related to Chat-GPT, it is argued that these three mechanisms (knowledge development, transparency of the artificial intelligence system, and ethical implications) have an important role in changing the speed and efficiency of processing big data to generate very accurate information. Therefore, the light was shed on the abovementioned mechanisms. as they provide a great insight into complex events by revealing hidden patterns and insightful information within data, facilitating improved data capture. The conclusion of the study led to the central role played by the ideal mechanisms of Chat-GPT in building a broad and reliable structure of information about various events and tasks, regardless of their complexity. Therefore, there is a good possibility to form a future research framework that about the role of ideal mechanisms of Chat-GPT in the metaverse applications.

Keywords: Chat-GPT · Metaverse · Production Management · Knowledge · Transparency · And Ethics

1 Introduction

Artificial intelligence, henceforth AI, technologies provide modern and advanced applications. Using these applications makes it easier for companies to accomplish tasks in several aspects, including collecting and analyzing data (Nasseef et al. 2022; Al-Abrow et al. 2021), accessing the required information in a very short time, making

M. Al-Emran et al. (Eds.): IMDC-IST 2024, LNNS 895, pp. 131–141, 2023.
https://doi.org/10.1007/978-3-031-51716-7_9

decisions based on filtered and well-studied information (Zhang et al. 2023), scheduling and preparing work tasks according to precise and specific timings, and providing accurate and improved information about the external environment, especially competitors. In addition, developing and implementing digital marketing strategies is another aspect (Tsai et al. 2023). One of the latest applications of AI at the present time is Chat-GPT, as it represents a common platform for various specializations in various companies (Dwivedi et al. 2023; Albahri et al. 2021). Product marketing systems usually require high-speed and accurate tools for obtaining information, classifying it, and selecting the best one. On the other hand, the metaverse is one of the modern tools used in marketing products as well (Buhalis et al. 2023). The spacious metaverse space provides rich experiences for users. The metaverse can make the user live in a virtual reality that explains to the user the essential or complementary characteristics of the product (Barrera and Shah 2023). Therefore, the need has grown to inject the ideal Chat-GPT mechanisms into metaverse applications, which has the ability to express reality through the accurate and transparent information it provides, along with the ethical aspect that studies the possibility of the user having a successful and honest experience with the company (Pellegrino et al. 2023). The delicate and focused study of the literature related to AI showed that there were no studies that experimented the integration of these two tools into the reality of companies' work. Although there is a lot of convergence between the aforementioned tools and the possibility of integration between them to achieve efficient use of the metaverse, it is possible to present an innovative proposal in the field of the integration process between two of the best AI applications. Also, through this proposed framework, the knowledge and application gap in this field can be bridged, as well as the possibility of expanding the horizons of future studies. Previous literature on AI has praised its potential to be integrated with many digital applications.

For example, a study (Estrada et al. 2023; AL-Fatlawey et al. 2021) addressed the possibility of applying Chat-GPT mechanisms to expand the knowledge characteristics of the consumer through the presentation process provided by metaverse applications. Thus, there is considerable value for Chat-GPT in strengthening the capabilities of other digital programs and shaping the features of their new capabilities. Also the study conducted by (Saddik and Ghaboura 2023) founded that there is a possibility to integrate between Chat-GPT and metaverse when apply them for medical consultations. Chat-GPT can also convey a better image to the consumer about the possibility of recording new characteristics of the product and clearly developing what will change in the characteristics of this product in the future (Rivas and Zhao 2023). Consequently, previous studies provided a view of the possibility of integration between these two tools. With the apparent scarcity of such studies. Especially now that the era is moving to digital applications very rapidly (Lv 2023). In addition to the great ability of the characteristics of these digital tools to eliminate obstacles resulting from the lack of knowledge and skill among working individuals (Huh 2023; Alnoor et al. 2022). Hence, we can say that the current study has highlighted the possibility of integration between Chat-GPT and the metaverse and has proposed a framework that can be adopted in the future. As a contribution, the study presented contributes promisingly to changing the features of marketing competition between companies. Policymakers in companies must also take into account the possibility of applying digital tools and combining them with each other, thus forming hybrid digital marketing strategies.

2 Literature Review

2.1 Knowledge Generation

Today, humans live in an era of generating highly accurate knowledge (Hill-Yardin et al., 2023). In past, generating knowledge required a lot of time, effort, financial, and human resources (Hu et al., 2023). Especially in the field of innovation and launching new products (Naqvi et al., 2023). Today, with the presence of artificial intelligence technologies, knowledge generation has exceeded the limits of human capabilities to an unlimited extent (Goodman et al., 2023; Alsalem et al., 2022; Atiyah, 2020). Thanks to the ability of artificial intelligence to process huge data quickly. As well as its ability to choose good or improved data.

2.2 Accelerated Data Processing

The rapid processing of data means the possibility of using modern and advanced methods and tools to convert data into valuable information (Bork and De Carlo, 2023). The best and most modern of these methods is artificial intelligence (Pietronudo et al., 2022). Previously, there were complex problems related to big data. But today, thanks to artificial intelligence, this problem has ended (Dwivedi et al., 2021). Therefore, the speed of data processing is an essential feature in today's world, which is called a paperless world (Singh et al., 2022). Therefore, data processing is achieved by adopting the following mechanisms: Data Collection, Data Ingestion, Data Preprocessing, Parallel Processing, Distributed Computing, In-Memory Processing, GPU Acceleration, Streaming Data Processing, Optimized Algorithms, Advanced Analytics, Scalability, Real-Time Insights, and Data Visualization.

2.3 Enhanced Data Exploration

Enhanced data exploration results of a deeper understanding of it (Zhang et al., 2023a, b). This process used to take a long time. With the large amount of data, it became impossible (Langford 2015). Artificial intelligence was capable of carrying out this task in an efficient way (Solaimani and Swaak 2023), by revealing hidden aspects and patterns in the data and being able to extract the best and most valuable ones. The ability of artificial intelligence to solve complex nodes in data is beyond reasonable limits (Farrokhi et al. 2020). Therefore, artificial intelligence has the ability to encode, encrypt, and decode the connections between data. And thus, reshaping that data as required by the context of the problem or situation (Burger et al. 2023). This is done by adopting a number of advanced mechanisms; as Data Understanding, Data Visualization, Interactive Dashboards, Exploratory Data Analysis (EDA), Dimensionality Reduction, Clustering and Classification, Feature Engineering, Time-Series Analysis, Advanced Analytics, Natural Language Processing (NLP), Geospatial Analysis, Anomaly Detection, Iterative Process, and Storytelling.

2.4 Transparency in AI Systems

Artificial intelligence has become a necessity for various specializations and situations of society (Cao et al. 2021) such as the field of health and environmental issues, marketing, individual behavior, and others (Bays et al. 2023; Polas et al. 2023; Gatea and Marina 2016). Therefore, transparency is an integral part of AI. It is very important to understand the results of artificial intelligence. The results are represented by a set of options presented by artificial intelligence, which are highly transparent (Singh et al. 2023). Transparency is embodied in verification, trust, and credibility. The ability to adopt complex concepts and present them in a clear and understandable way for all, depending on The Black Box Phenomenon and Interpretable.

2.5 The Black Box Phenomenon

The black box phenomenon indicates the great cognitive depth of artificial intelligence. This means that artificial intelligence adopts very complex models and applications in different contexts (Taylor et al. 2022). One of the most important features of these applications is the ambiguity in structuring the issues studied, filtering them, and taking the best aspects. This ambiguity indicates the great challenge these applications deal with. Its internal models cannot be interpreted by humans (Górriz et al. 2023). The black box phenomenon can be summarized as follows: Complex AI Models, Limited Interpretability, Decision-Making Opacity, Challenges in Debugging and Bias Detection, Ethical Concerns, Regulatory and Legal Implications.

2.6 Interpretable AI

One of the most important characteristics of artificial intelligence mechanisms is its capacity for explaining difficult issues. It is completely opposite to what came before it (The Black Box Phenomenon) in terms of name, but it differs in terms of content (Chou et al. 2022; Fadhil et al. 2021). Explanation relates to the results reported from AI processes. These results are analyzable, measurable and interpretable by humans. These results usually require a high level of transparency and not be limited to a narrow angle (Angelov et al. 2021; Atiyah and Zaidan 2022).

3   Ethical Considerations

To ensure that decisions and behaviors are in line with moral standards, respect for human rights, and society well-being, ethical considerations are a critical component of decision-making and behavior. They involve evaluating moral principles, values, and potential repercussions. These factors are crucial across a range of industries, including technology, healthcare, business, and research because they direct people and organizations toward morally righteous decisions (Moore et al. 2019; Atiyah 2023). Fairness, justice, and the highest standards of integrity in all human pursuits are promoted by ethical concerns, which act as a compass for navigating difficult problems and conundrums.

3.1 Data Privacy and Security

In Chat-GPT, data security and privacy are of utmost importance because this cutting-edge AI language model interacts with user data and generates responses. To keep users' trust and guard against any weaknesses, Chat-GPT must follow strict data protection policies. This requires making sure that user interactions and data are kept private, aren't being utilized for unauthorized purposes, and are being shielded from dangers outside the system.

3.2 Human-AI Collaboration

Collaboration between humans and artificial intelligence (AI) is a prime example of how human ingenuity and AI's computational capability can come together to transcend conventional bounds (Russell 2019; Khaw et al. 2022). By combining the data processing, pattern recognition, and automation capabilities of AI with human creativity, intuition, and empathy, this collaborative paradigm enables us to fully realize the potential of both (Smith 2021; Hamid et al. 2021; Manhal et al. 2023).

4   Relationship Between Chat-GPT and Metaverse

In order to clarify the relationship between Chat-GPT and the metaverse, we start from the three axes that are ideal Chat-GPT mechanisms in the reality of digital technologies. Let's first consider the knowledge development mechanism that Chat-GPT gives the metaverse (Dwivedi et al. 2023). Chat-GPT has a huge stock of knowledge. By possessing the quantity and quality of data supported by the continuous updating process (Kocoń et al. 2023). For example, at the present time, GPT-4 was released instead of GPT-3.5, the first of which is considered an upgraded version. The development results from the injection of an updated amount of new knowledge that contributes significantly to diversifying experiences and introducing new ways of dealing with the user based on the consultations provided by Chat-GPT (Feng et al. 2023). Chat-GPT gives the ability to deal with new information, compare multiple options, and choose the best one among them. It allows to browse a wide area of information provided by one click instead of searching, investigating, and studying to choose the best source of information (Johnson et al. 2023). Therefore, it contributes extensively to the development of metaverse knowledge. Transparency is Chat-GPT's second ideal mechanism. Chat-GPT can provide direct and irrefutable information by relying on up-to-date and expanded information (George 2023). It has been subjected to many tests for the information it provides. So, we can say that the transparency feature of Chat-GPT helps the metaverse expand transparent and unambiguous chat points that are multi-specialized at the same time. Also, quick and creative responses are very important in the metaverse, and that is what the transparency element in Chat-GPT provides (Abdullah et al. 2022). This activity immerses the user in a creative and transparent experience. The third and final ideal mechanism of Chat-GPT is the ethical aspect (Ray 2023). Before we start talking about the ethical feature, we must emphasize that merging the previous two mechanisms will give the metaverse user the ability to create more realistic virtual worlds and make it easier for the user to move within the scope of real and renewable information. If we consider GPT and the metaverse combined together, our perception and ability to navigate virtual worlds will likely undergo significant changes (Park and Kim 2022). We can create more realistic virtual worlds by using Chat-GPT to design intelligent beings with the ability to understand and respond to human speech. When this happens, companies will have more opportunities to use the metaverse as a marketing and sales tool, and more people will want to visit it (Bojic 2022). E-learning, therapy, and online travel are just some of the new applications that can be developed as a result of applying GPT to the metaverse. As the GPT and metaverse evolves, future innovations are sure to be even more exciting. However, the importance of the third mechanism (ethics) remains an important element in maintaining the metaverse application to be renewable, developed, and acceptable to the user. This requires companies to take into account legal considerations. To apply what is legally possible and stay away from what is the opposite (Ray, 2023). To apply what is legally possible and stay away from what is not possible. Thus, we suggest the following:

Proposition: knowledge development, transparency of the artificial intelligence system, and ethical implications have positive and significant influence on integrating of Chat-GPT and metaverse.

5 Methodology

Due to the abundance and overlap of information about both Chat-GPT and the metaverse, we conducted a mini-review on one of the important topics that is most applied in several fields. At the same time, this type of research methodology was adopted to achieve the goal of the study. Thus, obtaining more accurate information about the effect of integrating the ideal mechanisms of Chat-GPT when applied in the metaverse space. The mini-review approach supports the process of careful analysis of the different opinions about phenomena and adopting the best ones that are most in contact with reality. In order to comply with the requirements of this type of research methodologies, the conceptual framework was proposed to determine the relationship between Chat-GPT and metaverse applications.

6 Challenges and Future Directions

The journey of integrating Chat-GPT with metaverse entails both impressive advancement and a number of significant difficulties that require our focus. Even though Chat-GPT has demonstrated its mastery of natural language comprehension and creation, it is crucial to recognize and resolve the following issues in order to clear the way for their bright future.

1. Ethical Considerations: As Chat-GPT becomes more ingrained in our daily lives, ethical questions about the information it produces are raised. It is a delicate job to strike a balance between the right to free expression and prohibiting improper or malevolent use. The most important thing is to make sure Chat-GPT respects moral bounds while remaining a helpful tool.

2. Bias Mitigation: Chat-GPT is subject to biases present in its training data, like many AI models. Addressing these biases and making its responses fair.
3. User Privacy and Security: It is crucial to safeguard user privacy and secure data when interacting with Chat-GPT. Sensitive information must be protected with strong data encryption, user permission processes, and strict access controls.
4. Explainability and Transparency: Users should be able to deal with a new model of the developed metaverse, especially in new applications that are crucial like legal advice or healthcare.
5. Technical Restrictions: Despite significant advancements, in integrating process still has trouble processing nuanced or contextually complex interactions. Its capacity to deliver precise and context-aware replies has to be improved, and this is an area of active research.
6. Scaling Responsibly: The environmental effect of developing and implementing AI models increases as their size and complexity increase. It's critical for the advancement of sustainable AI by applicate and develop integration process.
7. Multimodal Capabilities: A promising area for future development is enhancing metaverse capacity to comprehend and produce material in a variety of modalities, including text, photos, and videos. As a result, interactions will be deeper and more engaging.
8. Personalization and customization: A growing difficulty is giving users the ability to tailor to their unique requirements and preferences. It's crucial to strike the proper balance between customization and moral usage.
9. User Education: It is the joint responsibility of developers, educators, and policymakers to guarantee that users possess the literacy needed to participate appropriately with and comprehend their potential and limitations.
10. Based on what we explained, in below the proposal framework (Fig. 1):

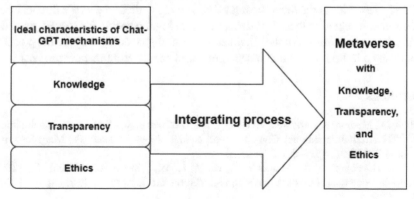

Fig. 1. Proposal Framework.

7 Challenges and Future Directions

Many companies have built their core and differentiated capabilities on the basis of significant developments in technology. Technology has made it possible to obtain a lot of knowledge in a short time and with very little effort. One of the fields of development in technology is AI. It represents a major tool in changing the work flow of companies in general. AI is involved in many areas of companies' work, whether in obtaining information, in the production process, or related to marketing and other activities. Therefore, companies began to compete greatly in getting the latest technology in the field of AI. Especially with the work of electronic platforms, which were considered a driving factor in the marketing of various products. Therefore, this study set out to describe the integration process between two tools that represent an applied model in the field of marketing. The importance of integration was that the metaverse takes into account knowledge development, transparency, and ethical considerations. Because the metaverse has become extremely influential in delivering information to the user. Depend on metaverse, the user lives in an experience that simulates tangible reality. Overcoming user doubts represents a company's success trait, and therefore the metaverse enables companies to reach that goal. The metaverse cannot continue to operate by relying solely on the company's goals. The metaverse cannot continue to operate by relying solely on the company's goals. The company may not take into account the level of knowledge that the user must obtain, or may not take into account the transparency of the information presented, or how the information is processed so that what is realistic and not imaginary reaches the user. As well as the importance of legal considerations at work. It is possible that while continuing to achieve the goals of companies, this application may be subject to cancellation or suspension by consumer protection organizations or governments. So, it is important to integrate these ideal mechanisms of Chat-GPT into the metaverse and thus obtain an honest visual and audio tool with advanced features, as well as high credibility and transparency in the information presented. Thus, we believe that our study presented a new idea at the theoretical level on how to address the user's doubts through the process of integrating these two tools, thus making it possible for specialists to delve into this field in the future. On the other hand, our study paved the way for the possibility of disclosing a hybrid tool that can be applied and benefit from its new properties.

References

Abdullah, M., Madain, A., Jararweh, Y.: ChatGPT: fundamentals, applications and social impacts. In: 2022 Ninth International Conference on Social Networks Analysis, Management and Security (SNAMS), pp. 1–8. IEEE (2022)

Al-Abrrow, H., Fayez, A. S., Abdullah, H., Khaw, K. W., Alnoor, A., Rexhepi, G.: Effect of open-mindedness and humble behavior on innovation: mediator role of learning. Int. J. Emerg. Mark. (2021)

Albahri, A.S., et al.: Based on the multi-assessment model: towards a new context of combining the artificial neural network and structural equation modelling: a review. Chaos Solitons Fractals **153**, 111445 (2021)

AL-Fatlawey, M.H., Brias, A.K., Atiyah, A.G.: The role of Strategic Behavior in achievement the Organizational Excellence" Analytical research of the manager's views of Ur State Company at Thi-Qar Governorate. J. Administration Econ. **10**(37) (2021)

Alnoor, A., Tiberius, V., Atiyah, A.G., Khaw, K.W., Yin, T.S., Chew, X., Abbas, S.: How positive and negative electronic word of mouth (eWOM) affects customers' intention to use social commerce? A dual-stage multi group-SEM and ANN analysis. Int. J. Hum.–Comput. Interact., 1–30 (2022)

Alsalem, M.A., et al.: Rise of multiattribute decision-making in combating COVID-19: a systematic review of the state-of-the-art literature. Int. J. Intell. Syst.Intell. Syst. 37(6), 3514–3624 (2022)

Angelov, P.P., Soares, E.A., Jiang, R., Arnold, N.I., Atkinson, P.M.: Explainable artificial intelligence: an analytical review. Wiley Interdisciplinary Rev. Data Mining Knowl. Discovery 11(5), e1424 (2021)

Atiyah, A.G.: The effect of the dimensions of strategic change on organizational performance level. PalArch's J. Archaeol. Egypt/Egyptology 17(8), 1269–1282 (2020)

Atiyah, A.G.: Strategic Network and Psychological Contract Breach: The Mediating Effect of Role Ambiguity

Atiyah, A.G., Zaidan, R.A.: Barriers to using social commerce. In: Artificial Neural Networks and Structural Equation Modeling: Marketing and Consumer Research Applications, pp. 115–130. Springer, Singapore (2022)

Barrera, K.G., Shah, D.: Marketing in the metaverse: conceptual understanding, framework, and research agenda. J. Bus. Res. 155, 113420 (2023)

Bays, H.E., et al.: Artificial intelligence and obesity management: an Obesity Medicine Association (OMA) Clinical Practice Statement (CPS) 2023. Obesity Pillars 6, 100065 (2023)

Bojic, L.: Metaverse through the prism of power and addiction: what will happen when the virtual world becomes more attractive than reality? Europ. J. Futures Res. 10(1), 1–24 (2022)

Bork, D., De Carlo, G.: An extended taxonomy of advanced information visualization and interaction in conceptual modeling. Data Knowl. Eng.Knowl. Eng. 147, 102209 (2023)

Buhalis, D., Leung, D., Lin, M.: Metaverse as a disruptive technology revolutionising tourism management and marketing. Tour. Manage. 97, 104724 (2023)

Burger, M., Nitsche, A.M., Arlinghaus, J.: Hybrid intelligence in procurement: disillusionment with AI's superiority? Comput. Ind.. Ind. 150, 103946 (2023)

Cao, G., Duan, Y., Edwards, J.S., Dwivedi, Y.K.: Understanding managers' attitudes and behavioral intentions towards using artificial intelligence for organizational decision-making. Technovation 106, 102312 (2021)

Chou, Y.L., Moreira, C., Bruza, P., Ouyang, C., Jorge, J.: Counterfactuals and causability in explainable artificial intelligence: theory, algorithms, and applications. Inf. Fusion 81, 59–83 (2022)

Dwivedi, Y.K., et al.: Artificial Intelligence (AI): Multidisciplinary perspectives on emerging challenges, opportunities, and agenda for research, practice and policy. Int. J. Inf. Manage. 57, 101994 (2021)

Dwivedi, Y.K., et al.: "So what if ChatGPT wrote it?" Multidisciplinary perspectives on opportunities, challenges and implications of generative conversational AI for research, practice and policy. Int. J. Inf. Manage. 71, 102642 (2023)

El Saddik, A., Ghaboura, S.: The Integration of ChatGPT with the Metaverse for Medical Consultations. IEEE Consumer Electronics Magazine (2023)

Estrada, J.S., Cruz, N.C., Lupión, M., Garzón, E. M., Ortigosa, P.M.: Artificial intelligence applied to teaching. In: EDULEARN23 Proceedings, pp. 2692–2701. IATED (2023)

Fadhil, S.S., Ismail, R., Alnoor, A.: The influence of soft skills on employability: a case study on technology industry sector in Malaysia. Interdiscip. J. Inf. Knowl. Manag.. J. Inf. Knowl. Manag. 16, 255 (2021)

Farrokhi, A., Shirazi, F., Hajli, N., Tajvidi, M.: Using artificial intelligence to detect crisis related to events: decision making in B2B by artificial intelligence. Ind. Mark. Manage. 91, 257–273 (2020)

Feng, Y., Vanam, S., Cherukupally, M., Zheng, W., Qiu, M., Chen, H.: Investigating code generation performance of chat-GPT with crowdsourcing social data. In: Proceedings of the 47th IEEE Computer Software and Applications Conference, pp. 1–10 (2023)

Franchi, S., Güzeldere, G. (eds.): Mechanical bodies, computational minds: artificial intelligence from automata to cyborgs. MIT press (2015)

Gatea, A.A., Marina, V.: Higher education funding in Iraq in terms of the experience of particular developed countries. Int. J. Adv. Stud. **6**(1), 8–17 (2016)

George, A.S., George, A.H.: A review of ChatGPT AI's impact on several business sectors. Partners Universal Int. Innov. J. **1**(1), 9–23 (2023)

Goodman, R.S., Patrinely, J.R., Osterman, T., Wheless, L., Johnson, D.B.: On the cusp: Considering the impact of artificial intelligence language models in healthcare. Med **4**(3), 139–140 (2023)

Górriz, J.M., et al.: Computational approaches to Explainable Artificial Intelligence: advances in theory, applications and trends. Inf. Fusion **100**, 101945 (2023)

Hamid, R.A., et al.: How smart is e-tourism? a systematic review of smart tourism recommendation system applying data management. Comput. Sci. Rev. **39**, 100337 (2021)

Hill-Yardin, E.L., Hutchinson, M.R., Laycock, R., Spencer, S.J.: A Chat (GPT) about the future of scientific publishing. Brain Behav. Immun.Behav. Immun. **110**, 152–154 (2023)

Hu, H., Li, R., Zhang, L.: Financial development and resources curse hypothesis: China's COVID-19 perspective of natural resources extraction. Resour. Policy. Policy **85**, 103965 (2023)

Huh, S.: Emergence of the metaverse and ChatGPT in journal publishing after the COVID-19 pandemic. Sci. Editing **10**(1), 1–4 (2023)

Johnson, D., et al.: Assessing the accuracy and reliability of AI-generated medical responses: an evaluation of the Chat-GPT model. Research square (2023)

Khaw, K.W., et al.: Modelling and evaluating trust in mobile commerce: a hybrid three stage Fuzzy Delphi, structural equation modeling, and neural network approach. Int. J. Hum.-Comput. Interact. **38**(16), 1529–1545 (2022)

Kocoń, J., et al.: ChatGPT: Jack of all trades, master of none. Information Fusion, 101861 (2023)

Langford, R.L.: Temporal merging of remote sensing data to enhance spectral regolith, lithological and alteration patterns for regional mineral exploration. Ore Geol. Rev. **68**, 14–29 (2015)

Lv, Z.: Generative Artificial Intelligence in the Metaverse Era. Cognitive Robotics. (2023)

Manhal, M., Al-khalidi, A., Hamad, Z.: Strategic network: managerial myopia point of view. Manage. Sci. Lett. **13**(3), 211–218 (2023)

Moore, C., et al.: Leaders matter morally: the role of ethical leadership in shaping employee moral cognition and misconduct, vol. 104, No. 1, p. 123. American Psychological Association (2019)

Naqvi, S.A.A., Hussain, B., Ali, S.: Evaluating the influence of biofuel and waste energy production on environmental degradation in APEC: Role of natural resources and financial development. J. Clean. Prod. **386**, 135790 (2023)

Nasseef, O.A., Baabdullah, A.M., Alalwan, A.A., Lal, B., Dwivedi, Y.K.: Artificial intelligence-based public healthcare systems: G2G knowledge-based exchange to enhance the decision-making process. Gov. Inf. Q. **39**(4), 101618 (2022)

Park, S.M., Kim, Y.G.: A metaverse: taxonomy, components, applications, and open challenges. IEEE Access **10**, 4209–4251 (2022)

Pellegrino, A., Stasi, A., Wang, R.: Exploring the intersection of sustainable consumption and the Metaverse: a review of current literature and future research directions. Heliyon (2023)

Pietronudo, M.C., Croidieu, G., Schiavone, F.: A solution looking for problems? a systematic literature review of the rationalizing influence of artificial intelligence on decision-making in innovation management. Technol. Forecast. Soc. Chang. **182**, 121828 (2022)

Polas, M.R.H., et al.: Artificial intelligence, blockchain technology, and risk-taking behavior in the 4.0 IR Metaverse Era: evidence from Bangladesh-based SMEs. J. Open Innov. Technol. Market Complexity **8**(3), 168 (2022)

Ray, P.P.: ChatGPT: A comprehensive review on background, applications, key challenges, bias, ethics, limitations and future scope. Internet of Things and Cyber-Physical Systems (2023)

Rivas, P., Zhao, L.: Marketing with chatgpt: Navigating the ethical terrain of gpt-based chatbot technology. AI 4(2), 375–384 (2023)

Russell, S.: Human compatible: Artificial intelligence and the problem of control. Penguin (2019)

Singh, A.V., et al.: Integrative toxicogenomics: advancing precision medicine and toxicology through artificial intelligence and OMICs technology. Biomed. Pharmacother.Pharmacother. 163, 114784 (2023)

Singh, J., Singh, G., Gahlawat, M., Prabha, C.: Big data as a service and application for indian banking sector. Procedia Comput. Sci. 215, 878–887 (2022)

Smith, B.C.: The promise of artificial intelligence: reckoning and judgment. Mit Press (2021)

Solaimani, S., Swaak, L.: Critical success factors in a multi-stage adoption of artificial intelligence. a necessary condition analysis. J. Eng. Tech. Manage. 69, 101760 (2023)

Taylor, A.H., Bastos, A.P., Brown, R.L., Allen, C.: The signature-testing approach to mapping biological and artificial intelligences. Trends Cogn. Sci.Cogn. Sci. 26(9), 738–750 (2022)

Tenbrunsel, A.E., Smith-Crowe, K.: 13 ethical decision making: where we've been and where we're going. Acad. Manag. Ann.Manag. Ann. 2(1), 545–607 (2008)

Tsai, M.L., Ong, C.W., Chen, C.L.: Exploring the use of large language models (LLMs) in chemical engineering education: building core course problem models with Chat-GPT. Educ. Chem. Eng. 44, 71–95 (2023)

Zhang, C., Zhu, W., Dai, J., Wu, Y., Chen, X.: Ethical impact of artificial intelligence in managerial accounting. Int. J. Account. Inf. Syst. 49, 100619 (2023)

Zhang, H., Zhu, L., Zhang, Q., Wang, Y., Song, A.: Online view enhancement for exploration inside medical volumetric data using virtual reality. Comput. Biol. Mcd.. Biol. Mcd. 163, 107217 (2023)

Use of Earnings Quality Measures to Evaluate Performance of Iraqi Banks in Terms of Industry 5.0: Using the Hierarchical Analysis Process

Zahraa Ali Al-Ridha Oudah[1], Jabir Hussein Ali[1(✉)], and Ahmed Ali Ahmed[2]

[1] Management Technical College, Southern Technical University, Basrah, Iraq
drjaber@stu.edu.iq
[2] Business Administration Department, Shatalarab University College, Basrah, Iraq
ahmedalrashed@sa-uc.edu.iq

Abstract. Industry 5.0 change the landscape of world. Several firms struggle to keep up with such technology. The concept of a "metaverse" in the context of banking or finance was not a well-established or widely discussed concept. However, it's possible that developments in the financial industry. This research recognizes employing measures of earnings quality as one of the tools for limiting profit management and one of the most important pillars in the process of attracting an investor to the shares of a corporation, given the transparency and credibility it provides in revealing an institution's financial conditions. In light of the multiplicity of measures of earnings quality, the disparity resulting from its employment may cause an investor confusion and inability to make an optimal investment decision. The research problem lies in how an integrated business model, where banks' performance is measured, can be built in order to help an investor make an investment decision in the stock market. The research also tries to answer how an investor can choose the optimal institutions in a particular sector given the multiplicity of earnings quality measures and to what extent do earnings quality standards (quality measures, performance evaluation measures and control measures) contribute to strengthening investment decisions. The research also seeks to extract new data by testing the aforementioned measures and their role in enhancing decisions and evaluating investment opportunities using the statistical program (Eviews v.9) and the hierarchical analysis process (AHP). The test was carried out on a sample of eight banks with shares listed on the Iraqi Stock Exchange for a five-year time series. The results showed a variation in the evaluation of the banks' performance of the research sample and a variation in the impact of measures of earnings quality on performance and investment decisions.

Keywords: Earnings quality · performance evaluation · investment decisions · hierarchical analysis process · metaverse

1 Introduction

Based on my most recent knowledge update in September 2021, it can be observed that the notion of a "metaverse" within the domain of banking or finance has not attained a firmly established or extensively deliberated status. However, it is conceivable that advancements in the financial sector have transpired subsequent to that period, potentially giving rise to novel terms or concepts. Within a broader framework, the mention of a "metaverse in banking" could perhaps allude to conjecture about the prospective integration of virtual or digital worlds within the domain of banking (Dwivedi et al., 2022). As an illustration, Virtual Branches: The utilisation of virtual reality or augmented reality technologies has the potential to establish virtual bank branches, enabling consumers to engage in transactions, seek guidance from bank employees, and get financial services through a more immersive and interactive medium (Muhsen et al., 2023; Al-Hchaimi et al., 2023; Chew et al., 2023). The increasing popularity of non-fungible tokens (NFTs) and blockchain technology has prompted banks to consider integrating these technologies into their services. This integration has the potential to establish digital ecosystems wherein customers can effectively manage their assets, including unique digital assets, within a virtual environment. Financial Education and Training: Banks have the potential to utilise virtual environments as a means to offer financial education, training, or simulations, with the aim of enhancing customers' comprehension and proficiency in the management of their financial affairs (Arpaci et al., 2022). Virtual banking advisers, also known as AI-powered banking advisors, have the potential to provide tailored financial guidance and services within a metaverse-like setting (Abbas et al., 2023; Bozanic et al., 2023). The utilisation of virtual spaces has the potential to bolster the security measures employed for safeguarding sensitive financial data by implementing sophisticated encryption and authentication techniques (Asif et al., 2023; AL-Fatlawey et al., 2021).

Since the emergence of the earnings quality, the identification of its issues and the completion of stages it has been through, academics gained great interest in the said concept since it challenges its exact definition. It is possible to identify situations in which quality evaluation becomes important and sheds light on its potential determinants. Studying earnings quality's concept in more depth, with increasing interest, does not mean studying the actual profit of the economic unit (Siems 2007) but rather studying the sustainable profit and the extent to which the unit is able to predict the expected profits in the future (Delkhosh et al. 2017; Gatea and Marina, 2016). In addition, it is related not only to the availability of net income but also to testing the ability of reported income to generate cash flows (Alhilfi and Ali, 2019). Its ability to achieve criticism has proven that the aforementioned traits are the cornerstones of its concept (Fantazy & Salem, 2014; Manhal et al., 2023). Therefore, the concepts of the earnings quality in the field of accounting vary according to the different users of the financial statements and their different viewpoints as to what the profits contained, making them enjoy quality (Atiyah and Zaidan, 2022).

2   Metaverse

Metaverse is a three-dimensional imaginary virtual world that resembles the real world and is controlled by concerned people through headsets, tactile gloves, avatars, and anthropomorphic characters that resemble the virtual world. The metaverse emerged in 1992 and adapted from the novel Shattering Snow by Neil Stephenson, where humans interact with each other as virtual characters in a three-dimensional virtual space similar to reality (Koohang et al., 2023; Alsalem et al., 2022). The metaverse was defined by

Mark Zuck as an integrated ecosystem between the virtual world and the real world are homogeneous and similar to reality, allowing the use of avatars and holograms to work and interact in the way the individual wants (Dwivedi et al., 2022; Albahri et al., 2021). The metaverse applications have been used in several fields, including marketing, education, health, social communication, tourism, entertainment, wars, design, and several other fields. Furthermore, during the previous pandemic of COVID-19 the using of metaverse has been increased in order to facilitate the work and overcome the issues of closure of many organizations because COVID-19 crisis led to change the managing of businesses and services and converted the work from face to face into virtual reality (Arpaci et al., 2022; Eneizan et al., 2019). The utilizing of metaverse has many benefits for countries, firms, and people. Hence, such a tool raises the experience of customers to check the products virtually. Thus, the metaverse affected the behavior of individuals positively and negatively, as it became a place for solving social problems and exchange of experiences between individuals from each other (Oh et al., 2023; Hamid et al., 2021; Fadhil et al., 2021). Despite the benefits achieved by the metaverse, it faces many risks that directly affect society in general and individuals. Many people in the virtual world are subjected to silvery and tactile harassment, abuse, bullying, hatred and racism, although it is an unreal reality, but it negatively affects the individual's psyche in the real world. For example, the BBC News pretended to be 13 years old, witnessed a threat of rape, racism and hatred in the world of Metaverse, and this negatively affects the real world and causes depression, psychological, tension and anxiety (Khaw et al., 2022; Alnoor et al., 2022). There is also a female doctor who was subjected to extortion and rape. There are several reasons behind the prevalence of forbidden behaviors in the world of Metaverse, perhaps online prohibition, addiction, mental illness, and lack of supervision and follow-up cause involvement in ethical behaviors in the virtual world (Cheah & Shimul, 2023; Al-Abrrow et al., 2021).

3   Methodology

It means discussing the integrated criteria for measuring the earnings quality represented by the measures (quality, performance evaluation, and control) that have been developed by researchers, and in different directions that are related to this concept, since there is no generally accepted way to measure the quality of profits, which relies on a series of measures that enhance the usefulness of the earnings figure for decision-making (Rodríguez & Gutiérrez, 2019), and is classified as follows:

$$\frac{TCAj, t}{Assetj, t-1} = \alpha 0 + \alpha 1 \frac{CFOj, t-1}{Assetj, t-1} + \alpha 2 \frac{CFOj, t}{Assetj, t-1} + \alpha 3 \frac{CFOj, t+1}{Assetj, t-1} 1 + \varepsilon j, t \tag{1}$$

The Accruals Quality: The gap between profits and cash is known as receivables (Ma & Ma, 2017); hence, the accruals quality is a measure of the earnings quality on the basis that the closer the profit is on the cash flows, the higher the quality. It has been found that the unit that has high receivables (any large gap between net income and operating cash flows) suffers from poor performance in the following year, and it is measured by the following equation (Annes et al., 2018).

$$X_{t+1} = \alpha + \beta X_t + \varepsilon_t \tag{2}$$

Earnings Persistence: This is the ability of profits to repeat itself, that is, profits result from the operating activities of the unit and not from temporary activities. It is more useful to users for the purposes of evaluation; therefore, persistence was used as a measure of the earnings quality (Francis et al., 2004). The state of sustainable revenue growth, the earnings persistence, is measured by the simple regression equation as follows (Francis et al., 2004).

Predictive Value: The ability to predict profits refers to the possibility of forecasting profits in a later period of time (Ma & Ma, 2017), and it is defined as the ability of profits to predict future profits or cash flows. It is a measure of the quality of profits. As more profits tend to be repeated, the higher the quality becomes (Frances et al., 2006). The ability to predict is calculated according to the following equation (Menicucci, 2020).

Return on Assets: This is one of the main measures economic unit seeks to achieve. This measure is important because it directly affects the financial risks the unit faces; hence, the greater the profits achieved by the unit, the lower the probability of its failure and the greater its ability to continue. Return on assets is measured according to the following equation (Ma & Ma, 2017).

$$\frac{EBIT}{assets} \tag{3}$$

The Ratio of Profits to Sales: This is an alternative measure of profitability that is measured according to the value of sales and is represented by profitability in terms of the sales activity of the economic unit. Its effect is proportional with the quality of profits, i.e. the higher the percentage of profits, the greater the earnings quality in the unit. It is measured according to the following equation: (Ma & Ma, 2017).

$$\frac{EBIT}{sales} \tag{4}$$

Measuring Non-Discretionary Accruals According to Total Assets and Operating Cash Flow: In order to obtain a performance measure that is relatively free from manipulation, the performance measure proposed by Cornett et al. (2008) is adopted. This reflects negatively on the quality of profits, and its effect is reversible, i.e. the higher the percentage of non-discretionary accruals, the lower the earnings quality in the economic unit. This measure is used as an indication of the level of profit reversal of the unit's economic reality, as it represents a measure of profit management that affects the level of validity of the profit reversal of the economic reality (Ma & Ma, 2017). It is calculated through the following equation:

$$CFO/assets + \%NDA \tag{5}$$

Financial Leverage: This is the degree of the unit's dependence in financing its assets on fixed income sources, e.g. loans, bonds or preferred stocks, which affects the profits that owners get as well as the degree of risk they are exposed to. Financial leverage is also defined as using other people's money (including loans or preferred shares that are amortized) with fixed financial costs, in which the economic unit must commit to paying it because of a greater risk it will face when the leverage ratio increases finance and vice versa (Smith & Premti, 2020). This may also lead to large fluctuations in earnings before tax (PBT), (Al-Shamaileh & Khanfar, 2014); hence, its increase leads to an increase in the debt ratio in economic units (Kanderová, 2019). Therefore, we can say that the increase in financial leverage has a negative effect on the earnings quality and vice versa. It is measured according to the following equation (Ma & Ma, 2017).

$$Leverage = \frac{liability}{assets} \tag{6}$$

Sales Growth: Growth means the economic unit operates continuously, which creates positive results, in the sense that higher sales growth will have an impact on increasing the unit value (Putri et al., 2020). There is a direct relationship between sales stability, company profits and debt financing; hence, the more stable the sales and profits of the economic unit, the more it benefits from the leverage of financing and the less risk it faces than if its sales were volatile. Sales growth is measured according to the following equation:

$$Growth of sales = \frac{salest - salest - 1}{salest - 1} \tag{7}$$

Information disclosure of Industry 5.0: This variable was measured using content analysis of banks reports. Several previous studies recommended this tool of measure such a concept in order to capture the information linked with metaverse in terms of industry 5.0 (Asif et al. 2023).

4 Results and Discussion

According to the study problem and the data available in the Iraq Stock Exchange, for the purpose of using the hierarchical analysis process in helping investors to make a rational investment decision, a checklist form was directed to a group of financial experts. The test results were as follows:

1) Average answers of respondents on the checklist:
2) Calculating the relative importance of the main criteria in the hierarchical analysis model by computing the following main comparison matrix

$$A = \begin{bmatrix} 1 & \frac{W1}{W2} & \cdots & \frac{W1}{Wn} \\ \frac{W2}{W1} & 1 & \cdots & \frac{W2}{Wn} \\ \frac{Wn}{W1} & \frac{Wn}{W2} & \cdots & 1 \end{bmatrix}$$

$$\begin{bmatrix} 1 & 1/7 & 1/7 \\ 7 & 1 & 1/7 \\ 7 & 1/7 & 1 \end{bmatrix} =$$

3) Calculate the normalized matrix

$$N = \begin{bmatrix} 1 & 0.142 & 0.142 \\ 7 & 1 & 0.142 \\ 7 & 0.142 & 1 \end{bmatrix}$$

$$= \begin{bmatrix} 0.067 & 0.111 & 0.111 \\ 0.467 & 0.778 & 0.111 \\ 0.467 & 0.111 & 0.778 \end{bmatrix}$$

4) Calculating the materiality weights for the three main criteria:

$$W = \begin{bmatrix} 0.10 \\ 0.45 \\ 0.45 \end{bmatrix}$$

5) The checklist matrix homogeneity test:

$$n_{max} = AW$$

$$= \begin{bmatrix} 0.10 \\ 0.44 \\ 0.44 \end{bmatrix} \begin{bmatrix} 0.10 \\ 0.45 \\ 0.45 \end{bmatrix} \begin{bmatrix} 0.067 & 0.111 & 0.111 \\ 0.467 & 0.778 & 0.111 \\ 0.467 & 0.111 & 0.778 \end{bmatrix}$$

$$n_{max} = 0.10 + 0.44 + 0.44 = 0.98$$

Through the five-year financial data (2013–2017) obtained from the banks, the financial ratios for the three basic standards adopted were calculated, leading to the computation of the relative importance of the sub-financial ratios. Table 1 is the decision matrix that includes the sub-relative importance of financial ratios. The decision matrix shows the relative importance of all sub-metrics represented by measures (quality of receivables, persistence, predictive value, profitability to sales, return on assets, non-discretionary accruals according to total assets, operating cash flow, financial leverage, information disclosure of industry 5.0 and sales growth). It was calculated according to Linear: Normalization Max method (Vafaei et al., 2016).

The hierarchical analysis process is a tool for multi-criteria decision-making that involves analyzing the quantitative and qualitative aspects of the decision, describing it as a scientific approach to formulating multi-criteria decision problems in the form of a structure. The preferences of each alternative according to each criterion and then translated into numerical values that make the selection and decision-making process more transparent (Atiyah, 2020). The limits of a specific scale (Saaty, 2008), and it is a useful tool in the event that the decision maker is unable to build the utility function on the basis of which he chooses the best alternative (Ishizaka & Nemery, 2013; Atiyah, 2023). For the purpose of selecting the best investment bank among the eight banks, a hierarchical analysis model for the study is employed for each bank separately.

Table 1. Decision Matrix Containing the Relative Importance of Financial Sub-Ratios.

Financial ratios	IC	SC	MC	NB	BC	II	AM	AS
Accruals quality	11.93	4.91	−11.13	2.37	−7.21	3.47	1.00	−4.86
Persistence	0.38	0.21	0.31	0.31	0.27	0.37	0.32	1.00
Predictive value	0.33	0.37	0.55	0.57	0.41	0.66	1.00	0.47
Return on assets	0.74	0.18	0.57	0.55	0.41	0.76	0.51	1.00
Profitability to sales	0.72	0.28	1.00	0.48	0.47	0.70	0.45	0.79
Non-optional benefits	−2.39	14.05	−1.04	−2.75	−9.18	3.93	1.00	12.88
Leverage	0.66	1.00	0.43	0.56	0.39	0.72	0.97	0.90
Sales growth	0.73	0.24	0.81	0.64	0.08	0.39	1.00	0.32
Industry 5.0	−0.01	0.04	0.06	1.00	0.11	0.58	−1.11	0.00

Notice: IC = Iraqi Commercial, SC = Sumer Commercial, MC = Mansour Commercial, NB = National Bank, BC = Baghdad Commercial, II = Iraqi Islamic, AM = Al-Mosul, AS = Ashur

The results of the respondents in the checklist showed that the criteria for performance evaluation and control are more important than the quality standard for the purpose of investment. Thus, the performance evaluation standard represented by measures (return on assets, profitability to sales and non-discretionary accruals) and the control standard represented by measures (financial leverage, growth of sales and growth of profits) have equal preference in the relative importance of investors and parties interested in investing. The quality criterion is still important despite it being less important than the standards of performance and control evaluation. This was demonstrated by calculating the weights of relative importance of the three main criteria. The reliability of the comparison matrix for the checklist proved to be consistent. The relative weight of the quality standard was (0.10), the relative weight of the performance evaluation standard was (0.45) and the relative weight of the control standard was (0.45); this proved that the relative weight of the performance and control evaluation criteria is equal (0.45) and that the quality standard is less important compared to the standards of performance and control evaluation. Table 2 shows the overall relative importance of each of the earnings quality criteria for the purpose of selecting alternatives (banks) according to preference in terms of investment. As shown in Table 2, this is done according to the order of relative importance of quality measures of total profits represented by the sum of quality measures, performance evaluation measures and control measures for each bank.

The result in Table 2 shows that Sumer Commercial Bank have the highest relative importance of total earnings quality criteria among the banks. Sumer Commercial Bank is followed by Ashur Commercial Bank, Iraqi Islamic Bank, Mosul Commercial

Table 2. The Relative Importance of Earnings Quality Criteria.

Rank	Banks	Quality	Performance	Control	Total measures
1	IC	1.26	−0.42	0.62	1.46
2	SC	0.55	6.53	0.58	7.65
3	AM	−1.03	0.24	0.58	−0.21
4	NB	0.33	−0.77	0.99	0.54
5	BC	−0.65	−3.74	0.26	−4.13
6	II	0.45	2.42	0.76	3.63
7	MC	0.23	0.88	0.39	1.50
8	AS	−0.34	6.60	0.54	6.81

Bank, Iraqi Commercial Bank, Iraqi National Bank, Al-Mansour Bank and Baghdad Commercial respectively. Alternatives (banks) according to preference is presented in Table 3.

Table 3. Banks in Order of Preference.

Rank	Alternatives (banks)	Relative importance of banks
1	Sumer Commercial Bank	7.654
2	Ashur Commercial Bank	6.807
3	Iraqi Islamic Bank	3.631
4	Mosul Commercial Bank	1.502
5	Iraqi Commercial Bank	1.464
6	Iraqi National Bank	0.542
7	Al-Mansour Commercial Bank	−0.208
8	Baghdad Commercial Bank	−4.130

In Table 3, the banks were identified and selected according to their relative importance for each of the quality measures, performance evaluation measures and control measures. They were arranged according to preference for the purpose of investment. Accordingly, it can be said that Sumer Commercial Bank achieves a higher investment opportunity for investors and parties interested in investment, followed by Ashur Commercial Bank and Iraqi Islamic Bank. The investment opportunities of these banks are high as a result of the high relative importance of the earnings quality in them, and the decisions are rational in these banks. However, in the Mosul and Iraqi Commercial Bank, the investment opportunities are medium because the relative importance of the earnings quality in it is medium and thus decisions become medium. As for the National Iraqi Bank, Al Mansour Bank and Baghdad Commercial Bank, investment opportunities are volatile due to the low relative importance of the quality of profits, and therefore decisions become unstable and random.

5 Conclusions

With regard to the relationship of measures to the quality of profits, the tests showed an inverse relationship between the earnings quality and the measure of the accruals quality, non-discretionary accruals and financial leverage. As for the rest of the measures represented by persistence, predictive value, return on assets, profitability from sales, sales growth and information disclosure of industry 5.0, they showed a positive relationship with the quality of earnings. The performance evaluation measures represented by the scale (return on assets, profitability to sales, non-discretionary accruals according to total assets and operating cash flow) are complementary measures of quality measures, as the results showed that most of the banks that enjoy high quality in terms of their quality measures adopt the performance appraisal metrics. The measures of control represented by measures (financial leverage, sales growth and information disclosure of industry 5.0) have proven their worth for the performance of the economic unit when evaluating financial position, through which manipulations and deviations in the economic unit are detected. This is what matters to investors and parties interested in the investment. Hence, the hierarchical analysis process accurately and objectively contributes to determining the best alternative. With the presence of multiple criteria and complex decisions, the analysis makes it easier to arrange the alternatives by showing the relative importance of each criterion. This helps the investors and interests' parties compare banks and identify which bank is in their interest to preserve and develop their rights within the money markets in Iraq.

6 Recommendations

The banking sector in the Iraq Stock Exchange field of study should be strengthened towards adopting earnings quality standards represented (quality measures, performance evaluation measures, and control measures) and its various sub-measures to get rid of deviations and fluctuations in profits. Important financial standards should be improved and the capabilities of the employees in the various departments should be developed through training courses, urging credibility of work, providing reliable data and cooperating with the departments (financial affairs, human resources, quality, research and development). It should also be ensured that the departments are administratively linked to one another. Long-term relationships with investors should be established. Investors should be actively involved in efforts to improve the earnings quality in the sector to obtain higher quality of profits. This will help the sector meet the requirements and expectations of investors as well as improve the response of investors and parties interested. Also, the banking sector should be encouraged to use earnings quality practices as they improve performance by reducing the level of profit management. From the practical side and the examination form, it can be observed that there is a disagreement in regarding the main earnings quality standards as equally important. However, the dimension of the metrics must be taken into consideration as there is a complete agreement among the financial experts on the importance of performance evaluation measures and control measures; they have equal preference in the importance of achieving the quality of profits, while the measures of the earnings quality are less important compared to the standards of performance and control evaluation. Given this preference, it is recommended to employ one of the criteria for evaluating performance or control when employing quality standards. This gives a clear vision for the institutions. To determine the earnings quality and to know the strengths and weaknesses of each bank, the earnings quality should have more than one measure. Thus, an evaluation of the financial situation and investment opportunities of the institutions is reached. The investors in the Iraq Stock Exchange should take into consideration the importance of the earnings quality to reinforce their decisions and serve as their guide to correct information. There is a need to use the hierarchical analysis process in selecting the best alternative. It is used in the case of multiple criteria and complex decisions to give accurate results and clear vision through the relative importance of each scale. Thus, the investment opportunities for investors are evaluated and the best alternative (bank) is determined according to the order of preference. Hierarchical analysis process is not only limited to the banking sector; it can also be used for any sector to meet the requirements and expectations of decision-makers as well as enhance the responsiveness of investors and interested parties honestly and accurately.

References

Abbas, S., et al.: Antecedents of trustworthiness of social commerce platforms: a case of rural communities using multi group SEM & MCDM methods. Electron. Commerce Res. Appl. **62**, 101322 (2023)

Al-Abrrow, H., Fayez, A.S., Abdullah, H., Khaw, K.W., Alnoor, A., Rexhepi, G.: Effect of open-mindedness and humble behavior on innovation: mediator role of learning. Int. J. Emerg. Market. (2021)

Albahri, A.S., et al.: Based on the multi-assessment model: towards a new context of combining the artificial neural network and structural equation modelling: a review. Chaos Solitons Fractals **153**, 111445 (2021)

Al-Fatlawey, M.H., Brias, A.K., Atiyah, A.G.: The role of strategic behavior in achievement the organizational excellence. Analytical research of the manager's views of Ur State Company at Thi-Qar Governorate. J. Administ. Econ. **10**(37) (2021)

Al-Hchaimi, A.A.J., Sulaiman, N.B., Mustafa, M.A.B., Mohtar, M.N.B., Hassan, S.L.B.M., Muhsen, Y.R.: A comprehensive evaluation approach for efficient countermeasure techniques against timing side-channel attack on MPSoC-based IoT using multi-criteria decision-making methods. Egypt. Inf. J. **24**(2), 351–364 (2023)

Alhilfi, M., Naeem, M.A., Ali, J.H.: Measuring the Earnings Quality in Iraqi Commercial Banks on the Basis of Net Income and Net Operating Cash Flows (2019)

Alnoor, A., et al.: How positive and negative electronic word of mouth (eWOM) affects customers' intention to use social commerce? A dual-stage multi group-SEM and ANN analysis. Int. J. Hum. Comput. Interact. 1–30 (2022)

Alsalem, M.A., et al.: Rise of multiattribute decision-making in combating COVID-19: a systematic review of the state-of-the-art literature. Int. J. Intell. Syst. **37**(6), 3514–3624 (2022)

Annes, M.C., et al.: The Relation between Earnings Quality and Corporate Performance for the Firms listed in the Lisbon Stock Exchange (2018)

Arpaci, I., Karatas, K., Kusci, I., Al-Emran, M.: Understanding the social sustainability of the Metaverse by integrating UTAUT2 and big five personality traits: a hybrid SEM-ANN approach. Technol. Soc. **71**, 102120 (2022)

Asif, M., Searcy, C., Castka, P.: ESG and Industry 5.0: the role of technologies in enhancing ESG disclosure. Technol. Forecast. Soc. Change **195**, 122806 (2023)

Atiyah, A.G.: The effect of the dimensions of strategic change on organizational performance level. PalArch's J. Archaeol. Egypt/Egyptol. **17**(8), 1269–1282 (2020)

Atiyah, A.G.: Strategic Network and Psychological Contract Breach: The Mediating Effect of Role Ambiguity (2023)

Atiyah, A.G., Zaidan, R.A.: Barriers to using social commerce. In Artificial Neural Networks and Structural Equation Modeling: Marketing and Consumer Research Applications, pp. 115–130. Springer, Singapore (2022)

Bozanic, D., Tešić, D., Puška, A., Štilić, A., Muhsen, Y.R.: Ranking challenges, risks and threats using fuzzy inference system. Decis. Making Appl. Manag. Eng. **6**(2), 933–947 (2023)

Cheah, I., Shimul, A.S.: Marketing in the metaverse: moving forward–what's next? J. Glob. Scholars Market. Sci. **33**(1), 1–10 (2023)

Chew, X., Khaw, K.W., Alnoor, A., Ferasso, M., Al Halbusi, H., Muhsen, Y.R.: Circular economy of medical waste: novel intelligent medical waste management framework based on extension linear Diophantine fuzzy FDOSM and neural network approach. Environ. Sci. Pollut. Res. 1–27 (2023)

Cornett, M.M., Marcus, A.J., Tehranian, H.: Corporate governance and pay-for-performance: the impact of earnings management. J. Finan. Econ. **87**(2), 357–373 (2008)

Delkhosh, M., Yakhdani, Z., Yadkhdani, Z.: The impact of conservatism on the earnings quality based on a moderating role of earnings: evidence from Tehran Stock Exchange (TSE). J. Econ. Manag. Perspect. **11**(4), 351–362 (2017)

Dwivedi, Y.K., et al.: Metaverse beyond the hype: multidisciplinary perspectives on emerging challenges, opportunities, and agenda for research, practice and policy. Int. J. Inf. Manage. **66**, 102542 (2022)

Eneizan, B., Mohammed, A.G., Alnoor, A., Alabboodi, A.S., Enaizan, O.: Customer acceptance of mobile marketing in Jordan: an extended UTAUT2 model with trust and risk factors. Int. J. Eng. Bus. Manag. **11**, 1847979019889484 (2019)

Fadhil, S.S., Ismail, R., Alnoor, A.: The influence of soft skills on employability: a case study on technology industry sector in Malaysia. Interdiscip. J. Inf. Knowl. Manag. **16**, 255 (2021)

Francis, J., LaFond, R., Olsson, P.M., Schipper, K.: Costs of equity and earnings attributes. Account. Rev. **79**(4), 967–1010 (2004)

Francis, J., Allen, H.H., Rajgopal, S., Zang, A.Y.: CEO Reputation and Earnings Quality (2006). www.ssrn.com

Gatea, A.A., Marina, V.: Higher education funding in Iraq in terms of the experience of particular developed countries. Int. J. Adv. Stud. **6**(1), 8–17 (2016)

Gutiérrez, A.L., Rodríguez, M.C.: A review on the multidimensional analysis of earnings quality. Rev. Contabil. Spanish Account. Rev. **22**(1), 41–60 (2019)

Hamid, R.A., et al.: How smart is e-tourism? A systematic review of smart tourism recommendation system applying data management. Comput. Sci. Rev. **39**, 100337 (2021)

Hoffmann, E.B., Malacrino, D.: Employment time and the cyclicality of earnings growth. J. Public Econ. **169**, 160–171 (2019)

Ishizaka, A., Nemery, P.: Multi-criteria Decision Analysis: Methods and Software. Wiley (2013)

Ištok, M., Kanderová, M.: Debt/asset ratio as evidence of profit-shifting behaviour in theSlovak Republic. Technol. Econ. Dev. Econ. **25**(6), 1293–1308 (2019)

Khaw, K.W., et al.: Modelling and evaluating trust in mobile commerce: a hybrid three stage Fuzzy Delphi, structural equation modeling, and neural network approach. Int. J. Hum. Comput. Interact. **38**(16), 1529–1545 (2022)

Koohang, A., et al.: Shaping the metaverse into reality: a holistic multidisciplinary understanding of opportunities, challenges, and avenues for future investigation. J. Comput. Inf. Syst. **63**(3), 735–765 (2023)

Ma, S., Ma, L.: The association of earnings quality with corporate performance. Pacific Account. Rev. (2017)

Manhal, M., Al-khalidi, A., Hamad, Z.: Strategic network: managerial myopia point of view. Manag. Sci. Lett. **13**(3), 211–218 (2023)

Menicucci, E.: IAS/IFRSs, accounting quality and earnings quality. In: Earnings Quality, pp. 83–105. Palgrave Pivot, Cham (2020)

Muhsen, Y.R., Husin, N.A., Zolkepli, M.B., Manshor, N., Al-Hchaimi, A.A.J. (2023). Evaluation of the Routing Algorithms for NoC-Based MPSoC: A Fuzzy Multi-Criteria Decision-Making Approach. IEEE Access

Oh, H.J., Kim, J., Chang, J.J., Park, N., Lee, S.: Social benefits of living in the metaverse: the relationships among social presence, supportive interaction, social self-efficacy, and feelings of loneliness. Comput. Hum. Behav. **139**, 107498 (2023)

Premti, A., Smith, G.: Earnings management in the pre-IPO process: biases and predictors. Res. Int. Bus. Financ. **52**, 101120 (2020)

Putri, I.G., Tri, A.P., Rahyuda, H.: Effect of capital structure and sales growth on firm value with profitability as mediation. Int. Res. J. Manag. IT Soc. Sci. **7**(1), 145–155 (2020)

Saaty, T.L.: Decision making with the analytic hierarchy process. Int. J. Serv. Sci. **1**(1), 83–98 (2008)

Salem, M.S.M., Fantazy, K.A.: The use of analytic hierarchy process (AHP) in the determination of earnings quality: the case of UAE. J. Mod. Account. Audit. **10**, 3 (2014)

Shamaileh, M.O., Khanfar, S.M.: The effect of the financial leverage on the profitability in the tourism companies (analytical study-tourism sector-Jordan). Bus. Econ. Res. **4**(2), 251 (2014)

Siems, M.M.: Convergence in Shareholder Law. Cambridge University Press (2007)

Vafaei, N., Ribeiro, R.A., Camarinha-Matos, L.M.: Normalization techniques for multi-criteria decision making: analytical hierarchy process case study. In: Camarinha-Matos, L.M., Falcão, A.J., Vafaei, N., Najdi, S. (eds.) Technological Innovation for Cyber-Physical Systems, DoCEIS 2016. IFIP Advances in Information and Communication Technology, vol. 470, pp. 261–269. Springer, Cham (2016). https://doi.org/10.1007/978-3-319-31165-4_26

The Moderating Role of Big Data Analytics Capabilities in the Relationship Between Supply Chain Management Practices and Sustainable Performance: A Conceptual Framework from a Metaverse Perspective

Susan Sabah Abdulameer[1,2](✉) and Yousif Munadhil Ibrahim[2,3]

[1] Business Administration Techniques Department, Management Technical College, Southern Technical University, Basra, Iraq
susansabah50@gmail.com
[2] Department of Business Administration, Basrah University College of Science and Technology, Basra 61004, Iraq
[3] Basra Oil Company, Ministry of Oil, Basra, Iraq

Abstract. The shift between companies from competition based on products to competition between supply chains has also forced companies to shift in the technology used in managing and sharing supply chain data. Therefore, the current study aims to create a conceptual framework on supply chain management practices (SCMP), big data analytics capabilities (BDAC), and sustainable performance (SP) from a metaverse perspective. The methodology used in the current conceptual study includes a wide review of the research literature and previous studies related to the research variables. On the basis of such review, model links were created and supported by the dynamic capability theory and the natural resource-based view (NRBV) theory. Accordingly, some implications were presented to academics and practitioners, and some future studies were also suggested.

Keywords: Supply Chain Management Practices · Sustainable Performance · Big Data Analytics Capabilities · Metaverse

1 Introduction

In the world today, the metaverse issue has become the latest trend in advanced technology. In the same context, integrating big data analytics capabilities (BDAC) as one of the metaverse applications into supply chain management practices (SCMP) is critical for achieving sustainability in companies. Several opportunities within metaverse applications exist for supply chain and operations management. One of the key advantages lies in the data-sharing aspect, integrated with extensive BDAC. For all stakeholders involved, this integration improves transparency, optimization, and collaboration in the supply chain processes (Dwivedi et al., 2022).

© The Author(s), under exclusive license to Springer Nature Switzerland AG 2023
M. Al-Emran et al. (Eds.): IMDC-IST 2024, LNNS 895, pp. 154–170, 2023.
https://doi.org/10.1007/978-3-031-51716-7_11

Supply chain management (SCM) is a holistic approach that commences with the strategic planning and supervision of the flow of logistics, information, materials, and services, extending from vendors to service providers or manufacturers, all the way to the end consumers (Jabbour, Filho, Viana, & Jabbour, 2011). This approach explains a crucial shift in the realm of business administration practices (Jabbour et al., 2011). SCM strikes are considered one of the most efficient actions for businesses to enhance their overall performance (Ou et al., 2010). To achieve the objective of optimizing supply chain operations for the sake of enhancing company performance, it is imperative to enhance the planning and management of various activities, including material planning, capacity planning, inventory control, and logistics, in collaboration with traders and customers (Chandra and Kumar, 2000; Albahri et al., 2021).

Within the supply chain framework, it is crucial to adopt management strategies that not only enhance the performance of the company and the broader supply chain but also prioritize social, economic, and environmental considerations (Govindan et al., 2014; Beske, 2012; Amin and Zhang, 2014; Al-Abrrow et al., 2021).

At present, key areas for research recommendations in supply chain management (SCM) encompass aspects such as coordinating the supply chain, distribution and transportation, inventory management, order processing, strategic planning and efficiency improvement, supply chain integration, supply chain data, reverse logistics, and information, as well as the selection of suppliers and vendors, and sustainable or environmentally conscious SCM (Dwivedi et al., 2022). Hence, the cultivation of a data-centric culture and the acquisition of analytical abilities will emerge as vital topics for future exploration. Additionally, there is a need to tackle issues related to equitable compensation for prosumers' intellectual property and other contributions, especially if Metaverses are envisioned as a lasting digital tool for organizations in the workplace (Dwivedi et al., 2022; AL-Fatlawey et al., 2021).

Companies are increasingly grappling with the challenges posed by the vast volume of data known as "Big Data," which, in the past few years, has become a thrilling area for efficiency and prospects (Akter, Wamba, Gunasekaran, Dubey, & Childe, 2016; Alnoor et al., 2022). The capability for big data analytics (BDA) is widely recognized as a transformative force in the way businesses operate (Akter et al., 2016; Davenport and Harris, 2007; Alsalem et al., 2022). Current literature underscores BDAC's potential to reshape both theory of management and practical applications, with references to it as "the potential to transform management theory and practice" (George et al., 2014, p. 325), it is the "next big thing in innovation" (Gobble, 2013, p. 64); and "the fourth paradigm of science" (Strawn, 2012, p. 34); or the next "management revolution" (McAfee and Brynjolfsson, 2012). The continuous expansion of global investments in BDAC remains unabated as companies seek to establish a lasting competitive edge. In 2013, these investments aimed at harnessing BDAC amounted to approximately $2.1 trillion (Lunden, 2013, and projections for 2014 indicate they will reach roughly $3.8 trillion (Gartner, 2014; Akter et al., 2016; Atiyah, 2020).

Furthermore, BDA use is regarded as a critical component of corporate performance in the worldwide market (Minelli et al., 2013; Wang & Hajli, 2017). Companies are currently having difficulty managing big data (BD) as a result of the fast-growing global data, data complexity, data privacy, etc. Additionally, because to growths in information and communication technology, for example the Internet of things and Web 2.0 (Waller & Fawcett, 2013; Abdul Moktadir et al., 2019), the volume of global data has grown significantly. As a result of this development, there are numerous chances to create BDA tools and use BD methods in manufacturing supply chains (Moktadir, Ali, Paul, & Shukla, 2019). Therefore, BDA may assist manufacturing supply chains in a variability of ways, including decision-making, the launch of new products, risk management and mitigation, market research for specific products, operational procedure improvement, and so forth (Abdul Moktadir et al., 2019; Schoenherr & Speier-Pero, 2015; Zhong et al., 2016).

Through the extensive systematic review presented b (Arunachalam, Kumar, & Kawalek, 2018), they emphasized the need to study the BDAC as a supporting and enhancing factor for SCMP. Moreover, based on a review of previous literature, there are numerous studies that have examined the relation between SCMP and performance (e.g., Govindan, Azevedo, Carvalho, & Cruz-Machado, 2014; Lenny Koh, Demirbag, Bayraktar, Tatoglu, & Zaim, 2007; Wook Kim, 2006). Moreover, there are many studies that have studied the correlation between BDAC and sustainable performance (SP) (e.g., Akter et al., 2016; Bag, Wood, Xu, Dhamija, & Kayikci, 2020a; Dubey et al., 2019). To the best of the researcher's knowledge and based on previous relevant studies, the moderating role of BDAC has not been highlighted. This represents the research gap that the current study will address. Therefore, this study aims to build a conceptual framework on the moderating effect of BDAC in the relationship between SCMP and SP from a metaverse perspective.

The next parts of the current study include: The second section is a review of literature and previous studies, followed by the third section, which involve the underpinning theories for the relations between the variables of the study. After that comes the fourth section, which contains the methodology used in the study. The last part concerns conclusions, limitations, and future research.

2   Literature Review

This section involves the intensive literature review on Metaverse in the SCM, SCMP along with both BDAC and SP in addition to underpinning theories of the relationships among these variables. A detailed description of these variables is provided in the following subsections.

2.1   Metaverse in Supply Chain Management

The metaverse is the upcoming technology revolution that will have an effect on society in the following decades through the production of immersive experiences in both real-world and virtual contexts (Buhalis, Leung, Lin, 2023). The real and virtual worlds are connected by the metaverse, which, albeit still hypothetical, allows for easy transitions between them (Buhalis et al., 2023). Lundmark (2022) explains "By 2030, every physical device that can be digitally connected will be. Eventually, every action in the digital world will have an effect on the physical world, and vice versa. So, the metaverse isn't dependent on a virtual reality headset. Rather, it's the coming together of complementary technologies, including cloud and edge (near the data source) computing, artificial intelligence, blockchain, the internet of things, virtual reality, augmented reality and digital twins." Dwivedi et al. (2022) suggest that "The metaverse has the potential to extend the physical world using augmented and virtual reality technologies allowing users to seamlessly interact within real and simulated environments using avatars and holograms. Virtual environments and immersive games (such as Second Life, Fortnite, Roblox and VRChat) have been described as antecedents of the metaverse and offer some insight to the potential socio-economic impact of a fully functional persistent cross platform metaverse".

Users of the Metaverse can communicate themselves more fully by using body language and hand gestures (Buhalis et al., 2023; Atiyah, 2023). It is commonly mentioned to as a personified Internet as a result, which can heighten presence and make online interactions more like in-person ones (Clegg, 2022; Fadhil et al., 2021; Gatea and Marina, 2016). The metaverse is consequently predictable to bring about a multitude of improvements and disturbances to every aspect of existence. This has effects on culture and society and presents markets and communities with both revolutionary opportunities and challenges around the globe (Dwivedi et al., 2023; Hamid et al., 2021).

Dolgui and Ivanov (2023). In their study, mentioned that additional metaverse advances may lead to the coexistence of physical supply chain and operations management (SCOM), metaverse SCOM, and SCOM for coordinating the physical and metaverse worlds. Moreover, in the SCOM, Queiroz et al. (2023) discovered significant parallels and divergences between metaverse adopters and non-adopters. As a result, various advantages and difficulties are anticipated both before and after adoption, but it is still pertinent, and there are some, though, that gain relevance after implementation. Therefore, the field of metaverse is very important in supply chain management, operations, and sustainability, and can have practical contributions in the near future.

2.2 Supply Chain Management Practices

According to Stevens (1989) and Charkha and Jaju (2014), a supply chain (SC) can be defined as a chain connecting numerous objects, from the supplier to the costumer, via both service and manufacturing, to handle the flow of information, goods, and money efficiently to satisfy business requirements. Researchers are now pushing for fundamental reforms in the way SCs have traditionally been run, with profit as the only goal (Govindan et al., 2014; Atiyah and Zaidan, 2022; Khaw et al., 2022). Due to the growing challenges posed by global warming and change of climate, efforts to make SCs more ecologically friendly have become more important (Shukla et al., 2009). The significance of tackling sustainability challenges in SCM has only lately been acknowledged by academics (Seuring and Müller, 2008; Gobbo et al., 2014).

SCM is defined by the "Council of Supply Chain Management Professionals (CSCMP)" as the management and coordination of all operations involving sourcing, conversion, procurement, and logistics administration, as well as interaction and cooperation with channel collaborators (Lenny Koh et al., 2007).

SCM, which aims to improve competitive performance, integrates a company's internal operations closely with its exterior operations—those of its suppliers, clients, and other channel members—and successfully connects them (Kim, 2006). SCMP must be successfully integrated with other departments within a company as well as externally with suppliers or consumers in order to attain improved supply chain performance (Narasimhan, 1997).

SCMP are a group of steps performed by a business to support effective SCM (Lenny Koh et al., 2007; Li, Rao, Ragu-Nathan, & Ragu-Nathan, 2005; Manhal et al., 2023); as techniques for integrating, managing, and coordinating connections, demand, and supply in order to successfully please clients (Wong et al., 2005); as tangible actions or technological advancements that have a substantial impact on how a focus firm works with its suppliers and/or clients. In order to connect the downstream processes, it consequently entails creating client contacts through customer feedback and providing orders in a straight line to clients (Chow et al., 2008). In this way, researching SCMP lends credibility to the SCM perspective theory (Jabbour et al., 2011).

"Supply chain practices" cannot improve their proficiencies on their own since efficiency can only be obtained via the interplay of several supply chain practices. According to Dawe (1994), in order for SCM to be effective, an organization's entire functions of supply chain must be improved. Furthermore, the emphasis of supply chain practices must be modified from being functional and independent to being general and integrative. This suggests that each "supply chain practice's" efficacy should be evaluated considering how it significantly affects the efficient integration of all supply chain operations (Wook Kim, 2006). In order to successfully achieve SC integration, numerous supply chain practices might be applied methodically. Ballou's (1992) finding that all supply chain procedures result in the primary activities of the chain has been used to bolster Dawe's (1994) assertions. These key actions play crucial roles in the efficient modification and linking of various supply chain functions (Kim, 2006). A company may take a number of steps known as SCMPs to promote effective SCM. There is a substantial corpus of study on the many elements of SCMP (Koh et al., 2007). Moreover, Chopra and Meindl (2001) mentioned to SC contains a number of tactics and actions to efficiently participate providers, traders, manufacturers, and clients for refining the performance of the companies and the SC in an interconnected and high-performing business model.

The previous ten years, the stages of the SC have seen the most digital revolution; as a result, many IoT apps have been developed to track and trace the manufacturing of raw materials (Kache & Seuring, 2017). Smart apps for the SC, which first appear with the physical practice before shifting to the digital, give the backdrop of interconnection among the many processes and stakeholders (Ivanov & Dolgui, 2020). According to Fantini et al. (2018), the metaverse practice moves the physical-digital balance while offering a digital setting in which we can recreate the physical supply chain stages. Thus, the manufacturing, purchasing, and warehouse management operations are just a few of the supply chain steps that the metaverse affects in many ways (Dwivedi et al., 2022).

2.3 Big Data Analytics Capabilities

Kauffman et al., (2012, p. 85), mentioned to the notion of big data is exploding "due to social networking, the internet, mobile telephony and all kinds of new technologies that create and capture data". Actually, according to Akter et al. (2016), businesses are buried in a sea of data that is mostly made up of transaction data (which includes organized information about commerce operations, client profiles), clickstream information (e.g.

as text about web pages and social network websites such as blogs, tweets, and postings of Facebook), movie content (e.g. as images from commerce and other types of stores), and voice information (e.g. telephone calls information, call centers, and service of customer). Goes (2014) claims that big data consists of vast quantities of observational data that can be used to support numerous conclusions.

Big data is an exciting new opportunity for prospective advantages related to social and economic value, also a potential basis of competitive advantage for establishments long-term and medium-term (Grover et al., 2018). Some academics claim that the distinctive qualities of "big data" can be summed up as "(i) volume; (ii) diversity; (iii) velocity; (iv) veracity; (v) variability; and (vi) value" (Gandomi & Haider, 2015). Big data's first characteristic is its sheer amount, and its second is the variety of its structural make-up. The third feature has to do with how quickly it is generated (Albergaria and Jabbour 2020). The intrinsic undependability of some major data sources is discussed in the fourth attribute, while the volatility in flow rates and the abundance of data sources in these situations are discussed in the fifth attribute (Albergaria and Jabbour 2020). The potential value that could be obtained from a thorough investigation of big data is the subject of the sixth attribute (Gandomi & Haider, 2015; Martens, Provost, & Clark, 2016).

BDA, on the other hand, refers to the "use of statistical, processing, and analytics techniques to big data" (Ferraris et al., 2019). According to one definition, BDA is "… a holistic process that involves the collection, analysis, use, and interpretation of data for various functional divisions with a view to gaining actionable insights, creating business value, and establishing competitive advantage…" (Akter et al., 2016). There are numerous examples of BDA being used in various contexts today, including management of business process (Dezi et al., 2018), company performance (Müller et al., 2018), creation of value in SCM (Dubey et al., 2019), organizational performance (Müller et al., 2018), and service innovation (Lehrer et al., 2018), only to name a few. According to Kiron et al. (2014) and Akter et al. (2016), BDAC is the "ability to give business insights through data management, infrastructure (technology), and talent (people) capabilities to transform business into a competitive force".

2.4 Sustainable Performance

The topic of sustainable development is divisive and has a wide range of opinions and attitudes (Shahbaz et al., 2021). The word is essentially defined as the point at which the economy, ecology, and society converge. The Brundtland report (Vachon and Mao, 2008), in particular, has drawn significant attention to the idea of sustainable development in recent decades. Business's role in sustainability is frequently seen as a "responsibility" to the social order, and this responsibility is described as the requirement to eradicate business's harmful effects (Giddings et al., 2002). If opportunities connected to sustainability can be properly identified, it can be a key to competitiveness (Ibrahim, Hami, Othman, 2019). Though SSCM and logistics management are the subjects that have been studied the most in the literature on sustainable development (Baumgartner, 2014), there is a strong academic need to ascertain whether SCMP have an influence on organizational performance (Zhu et al., 2022).

3 Underpinning Theory

3.1 Natural Resource-Based View (NRBV) Theory

The roots of the resource-based view (RBV) theory are mentioned on "Penrose's (2009)" as a seminal contribute, through his book entitled "The theory of the growth of the firm", has been stated to view businesses as an all-encompassing collection of resources that can be controlled, reorganized, and used to support the company's unique values (Dierickx & Cool, 1989; Barney, 1991). Wernerfelt (1984) focused on how organizations acquire competitive advantages via organizational procedures, actions, and rules by using both intangible and tangible resources and capabilities (Hitt, Xu, & Carnes, 2016; Grant, 1991). He did this by providing an explanation of the RBV theory. The company's physical assets, financial assets, human capital, organizational (social) procedures, and other assets are all considered resources (Hart & Dowell, 2011). The resource must be valued, uncommon, unrivalled, and supported by organizational (social) processes to generate sustainable competitive advantages (Barney, Wright, & Ketchen, 2001). Contrarily, the ability is clear as anything the business can perform because of the resources and practices it uses (Karim & Mitchell, 2000; Winter, 2000).

In a seminal article, Hart (1995) stated that RBV theory had a severe weakness in ignoring the company's interaction with its natural environment (Hoskisson, Wan, Yiu, & Hitt, 1999). This is because the natural environment can create major constraints when companies attempt to make sustainable competitive advantages (Hart & Dowell, 2011). Or, more positively, "it is likely that strategy and competitive advantage in the coming years will be rooted in capabilities that facilitate environmentally sustainable economic activity—a natural-resource-based view of the firm" (Hart, 1995, p. 991).

In Hart's (1995) analysis, the three primary strategic competencies of the NRBV theory are "pollution prevention, product stewardship, and sustainable development". Each of these numerous environmental driving forces is reliant on various important resources and derives its competitive edge from multiple sources (Hart & Dowell, 2011). "Pollution prevention" capabilities focus on the environmental efficiency of processes and products through reducing emissions and waste from these operations and products (Hart & Milstein, 2003). Product stewardship focuses on expanding pollution prevention along the value chain of the company's product process (Hart & Dowell, 2011). "Sustainable development" capabilities have a distinct advantage over pollution prevention capabilities and product stewardship, it is not limited to environmental aspects, butinclude interesting on economic and social aspects (Hart, 1995; Hart & Milstein, 2003).

The NRBV theory is one of the greatest vital contemporary theories applied in studies associated to environmental and sustainable processes and practices to improve SP (Bansal & Roth, 2000; Nidumolu, Prahalad, & Rangaswami, 2009; Sarkis, Gonzalez-Torre, & Adenso-Diaz, 2010; Sroufe, 2003; Wu, Melnyk, & Calantone, 2008). In this context, several studies dealing with topics of sustainable operations and practices have used NRBV theory as an underpinning theory (e.g., Aboelmaged, 2018; Huo, Gu, & Wang, 2019; Menguc & Ozanne, 2005). In light of this, NRBV theory addresses the connection between SCMP and SP.

3.2 Dynamic Capability Theory

As stated by the idea of dynamic capability, the firm can embrace constant change and maintain performance in a sector that is changing quickly. Dynamic capabilities are defined as "an organization's ability). So, BDAC is a dynamic capacity that brings together, integrates, and provides resources that are tailored for big data (Olszak, 2014). Establishment's decision-making and long-term performance may be enhanced by comprehending BDAC and utilizing it effectively (Zhu et al., 2022). As a result, this hypothesis supports both the moderating effect of BDAC as well as the relationship between BDAC and SP (Zhu et al., 2022).

In sum, this study suggests that SCMP have a positive effect on SP based on the NRBV theory's presumptions and the evidence provided. Additionally, the study finds a moderating role of BDAC on the link between SCMP and SP by using dynamic capability theory. Therefore, this study combines NRBV theory and dynamic capability theory to covering the research model.

4 Hypothesis Development

Figure 1 presents the research framework created in this study from a metaverse perspective. According to the framework, SCMP will have an impact on SP and will also have BDAC's assistance as a moderating factor to have a greater impact on SP from a metaverse perspective. The hypothesized links between SCMP, BDAC, and SP are addressed in light of the existing research, and theories pertaining to these variables are created.

4.1 Relationship Between Supply Chain Management and Sustainable Performance

There is evidence in literature about the influence of SCMP on SP. According to Govindan et al.'s (2014) research of five case studies from the Portuguese automobile supply chain, SCMP has a positive impact on SP. In the same vein, Koh et al., (2007) find SCMP have effect on performance of manufacturing SMEs in Turkey. Also, Kim (2006), founded in his study in small and large manufacturing firms SCMP and competition capability have more important impact on performance. Thus, based on the assumptions of the NRBV theory and the evidence presented, the current study proposes the following proposition:

Proposition 1: There is significant and positive association between supply chain management practices and sustainable performance.

4.2 Relationship Between Big Data Analytics Capabilities and Sustainable Performance

According to Wixom et al. (2013), BDAC is widely recognized to be essential for improving business and firm performance. Price optimization and profit maximization are two areas where the literature shows a correlation between BDAC and performance (Schroeck et al., 2012; Akter et al., 2016).

Big data has been referred to as a manufacturing environment transformation because to its enormous revolutionary powers in business, management, and research (Junaid et al., 2022; Schroeck et al., 2012; Ataseven & Nair, 2017; Kumar et al., 2017). To enable data-driven decision-making, using sophisticated analytics techniques, BDA is a means of pulling relevant information out of massive volumes of data (Nilsson & Göransson, 2021). The BDAC notion, in contrast, is definite as "the ability of an organization to integrate, build, and reconfigure the information resources, as well as business processes, to address rapidly changing environments" (Lee & Kang, 2015). Additionally, BDAC is a comprehensive, complementary capacity that aids businesses in effectively coordinative and using their information, knowledge, and technologies in order to collaboratively expand their present models and value-added processes (Shamim et al., 2020; Gharajeh, 2018). Thus, this research suggests the following in light of the evidence provided and the underlying assumptions of the dynamic capability theory:

Proposition 2: There is significant and positive association between big data analytics capabilities and sustainable performance.

4.3   The Moderating Effect of Big Data Analytics Capabilities

With the use of AI and multimodal data analytics, supply chain managers will have access to a variety of data on the various stages of the supply chain in the metaverse in order to manage it as efficiently as possible (Dwivedi et al., 2022; Sharma & Giannakos, 2020). The prior literature has gone into detail about the impact of BDA on SSCP. Several studies have emphasized the significance of BDA offers useful data for internal and external supply chain integration operations (Belhadi et al. 2020; Corbett, 2018; Kamble et al. 2020). According to studies by Belhadi et al. 2020, Mangla et al. 2020, and Liu, Zhu, and Seuring (2017), BDA can improve supply chain efficiency and help create more sustainable systems. Environmental impact assessments produced by BDA are more accurate and aid in the analysis of the impacts of SSCM (Kamble and Gunasekaran 2020). Through cost savings, enhanced coordination, and collaboration across the many supply chain players, a data-driven SSC helps businesses achieve their financial goals (Tseng et al. 2019). To maximize resource utilization in a BDA context, the decision-making support system is crucial (Gupta et al., 2019). Bag et al. (2020b) examined how BDAC affects SC performance and discovered that the management skills of the analytics team have a beneficial impact on environmental products, innovativeness, and SSC advantages. BDA is a crucial tool for tracking and monitoring social elements in a multi-level supply chain (Venkatesh et al. 2020). Barbeito-Caamao and Chalmeta (2020) claim that businesses can use BDA to deliver a further specialized and targeted response to an intended audience or issues and, finally, increase the value for the people by managing the data and insights from numerous stakeholders gathered from online social networks. Khan et al. (2021) claim that businesses might apply BDA methodologies to unlock, model, and forecast the behavior of social stakeholders (such as NGO's, employees, and public institutions), allowing the detection of their trends, and possibly arising social matters. Once the social matters are identified, this encourages collective solutions fit for the various stakeholders and makes more transformative and participative options available to fight back.

A number of variables, the emergence of the big data era was influenced by a number of factors, including an increase in the usage of supply chain technologies, an influx of data, and a change in management emphasis from heuristics to data-driven decision making (Arunachalam et al., 2018). BDA is a useful tool for making decisions and can have a big impact on changing and enhancing supply chain procedures. Corporate leaders prefer to base their decisions in this evolving business climate on data-driven visions rather than their gut feelings (Davenport, 2006). Due to the alleged advantages of BDA, organizations are very motivated to improve their technological and organizational capabilities to get value from data (Arunachalam et al., 2018). Over the past ten years, there has been a substantial increase in the use of various "Information and Communication Technologies (ICT)" for SCM, including the (IoT), RFID, and ERP. The supply chain has generated a ton of data because of this. The new era of big data is the consequence of our ongoing efforts to develop more sophisticated technology to capture data at various supply chain stages (Arunachalam et al., 2018). Based on the assumptions of the dynamic capacity theory and the presented facts, this study proposes that BDAC have a moderating influence on the link between SCMP and SP:

Proposition 3: Big data analytics capabilities have significant and positive moderate influence on the relationship between supply chain management practices and sustainable performance.

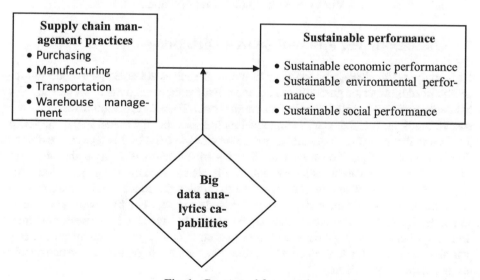

Fig. 1. Conceptual framework

5 Methodology

The current conceptual study used the approach of reviewing previous studies and literature extensively in order to build a conceptual framework that consists of three variables. The first, independent variable is SCMP, which includes the dimensions: manufacturing, purchasing, transportation, and warehouse management. The second, dependent variable is SP, which consists of the dimensions: "sustainable economic performance, sustainable environmental performance, and sustainable social performance". The last, moderating variable is BDAC. Natural resource-based theory was applied to the relationship between SCMP and SP in order to determine the correlations between these three variables. The dynamic capability theory covers the association between BDAC and SP, in addition to the moderating role of BDAC.

6 Conclusion

The purpose of this study is to build a conceptual framework, so an extensive review of previous studies and literature was conducted to achieve this aim. The study established a conceptual framework that includes three variables: SCMP, BDAC, and SP. This model included the proposed relationships, which involve the direct and positive association between SCMP and SP, as well as the direct positive influence relation between BDAC and SP. Also, the moderating role of BDAC. These relationships confirm the expected responses from the current study. These results can contribute appropriately to researchers specializing in supply chains, big data, and sustainability, by examining the nature of the relationships discovered in this study empirically. Practitioners can also be guided by these relationships when making relevant decisions.

7 Limitations and Future Research Directions

This study, despite its results, has some limitations, such as not testing the model empirically. In addition to the possibility of adding other mediators and moderators' variables related to metaverse applications in the current model. Therefore, the future research suggested to researchers that can be conducted includes testing the model in a quantitative method in the manufacturing industry or the service industry. In this regard, possible to focus on the oil and gas industry in quantitative testing because it is among the industries that need to use big data in its operations, which can enhance the implementation of its supply chain more efficiently. It is important to employ the concept of the metaverse in the variables of the present study and the proposed variables. The framework can also be expanded by adding some other variables. For example, given the importance of culture in the workplace, the organizational culture variable will be an important moderating variable in the model when added. In addition, the variable of SCMP can be considered as an independent variable.

References

Aboelmaged, M.: The drivers of sustainable manufacturing practices in Egyptian SMEs and their impact on competitive capabilities: a PLS-SEM model. J. Clean. Prod. **175**, 207–221 (2018). https://doi.org/10.1016/j.jclepro.2017.12.053
Akter, S., Wamba, S.F., Gunasekaran, A., Dubey, R., Childe, S.J.: How to improve firm performance using big data analytics capability and business strategy alignment? Int. J. Prod. Econ. **182**, 113–131 (2016). https://doi.org/10.1016/j.ijpe.2016.08.018

Al-Abrrow, H., Fayez, A.S., Abdullah, H., Khaw, K.W., Alnoor, A., Rexhepi, G.: Effect of open-mindedness and humble behavior on innovation: mediator role of learning. Int. J. Emerg. Market. (2021)

Albahri, A.S., et al.: Based on the multi-assessment model: towards a new context of combining the artificial neural network and structural equation modelling: a review. Chaos Solitons Fractals **153**, 111445 (2021)

Albergaria, M., Jabbour, C.J.C.: The role of big data analytics capabilities (BDAC) in understanding the challenges of service information and operations management in the sharing economy: evidence of peer effects in libraries. Int. J. Inf. Manag. **51**, 102023 (2020)

AL-Fatlawey, M.H., Brias, A.K., Atiyah, A.G.: The role of Strategic Behavior in achievement the Organizational Excellence "Analytical research of the manager's views of Ur State Company at Thi-Qar Governorate". J. Administ. Econ. **10**(37) (2021)

Alnoor, A., et al.: How positive and negative electronic word of mouth (eWOM) affects customers' intention to use social commerce? A dual-stage multi group-SEM and ANN analysis. Int. J. Hum. Comput. Interact. 1–30 (2022)

Alsalem, M.A., et al.: Rise of multiattribute decision-making in combating COVID-19: a systematic review of the state-of-the-art literature. Int. J. Intell. Syst. **37**(6), 3514–3624 (2022)

Amin, S.H., Zhang, G.: Closed-loop supply chain network configuration by a multi-objective mathematical model. Int. J. Bus. Perform. Supply Chain Model. **6**(1), 1–15 (2014)

Arunachalam, D., Kumar, N., Kawalek, J.P.: Understanding big data analytics capabilities in supply chain management: unravelling the issues, challenges and implications for practice. Transport. Res. Part E: Logist. Transport. Rev. **114**, 416–436 (2018). https://doi.org/10.1016/j.tre.2017.04.001

Ataseven, C., Nair, A.: Assessment of supply chain integration and performance relationships: a meta-analytic investigation of the literature. Int. J. Prod. Econ. **185**, 252–265 (2017)

Atiyah, A.G.: The effect of the dimensions of strategic change on organizational performance level. PalArch's J. Archaeol. Egypt/Egyptol. **17**(8), 1269–1282 (2020)

Atiyah, A.G. Strategic Network and Psychological Contract Breach: The Mediating Effect of Role Ambiguity (2023)

Atiyah, A.G., Zaidan, R.A.: Barriers to using social commerce. In: Artificial Neural Networks and Structural Equation Modeling: Marketing and Consumer Research Applications, pp. 115–130. Springer Nature Singapore, Singapore (2020)

Bag, S., Wood, L.C., Xu, L., Dhamija, P., Kayikci, Y.: Big data analytics as an operational excellence approach to enhance sustainable supply chain performance. Resour. Conserv. Recycl. **153**, 104559 (2020). https://doi.org/10.1016/j.resconrec.2019.104559

Bag, S., Yadav, G., Wood, L.C., Dhamija, P., Joshi, S.: Industry 4.0 and the circular economy: resource melioration in logistics. Resources Policy **68** (2020)

Ballou, R.A.: Business Logistics Management. Prentice Hall (1992)

Bansal, P., Roth, K.: Why companies go green: a model of ecological responsiveness. Acad. Manag. J. **43**(4), 717–736 (2000). https://doi.org/10.5465/1556363

Barbeito-Caamaño, A., Chalmeta, R.: Using big data to evaluate corporate social responsibility and sustainable development practices. Corp. Soc. Responsib. Environ. Manag. **27**(6), 2831–2848 (2020)

Barney, J.: Firm resources and sustained competitive advantage. J. Manag. **17**(1), 99–120 (1991). https://doi.org/10.1177/014920639101700108

Barney, J., Wright, M., Ketchen, D.J.: The resource-based view of the firm: ten years after 1991. J. Manag. **27**(6), 625–641 (2001). https://doi.org/10.1177/014920630102700601

Baumgartner, R.J.: Managing corporate sustainability and CSR: a conceptual framework combining values, strategies, and instruments contributing to sustainable development. Corp. Soc. Responsib. Environ. Manag. **21**, 258–271 (2014)

Belhadi, A., Kamble, S.S., Zkik, K., Cherrafi, A., Touriki, F.E.: The integrated effect of big data analytics, lean six sigma, and green manufacturing on the environmental performance of manufacturing companies: the case of North Africa. J. Clean. Prod. **252**, 119903 (2020)

Beske, P.: NOFOMA dynamic capabilities and sustainable supply chain management. Int. J. Phys. Distrib. Logist. Manag. **42**(4), 5–25 (2012)

Buhalis, D., Leung, D., Lin, M.: Metaverse as a disruptive technology revolutionising tourism management and marketing. Tour. Manage. **97**, 104724 (2023)

Chandra, C., Kumar, S.: Supply chain management in theory and practice: a passing fad or a fundamental change? Indust. Manag. Data Syst. **100**(3), 100–113 (2000)

Charkha, P.G., Jaju, S.B.: Supply chain performance measurement system: an overview. Int. J. Bus. Perform. Supply Chain Model. **6**(1), 40–60 (2014)

Chopra, S., Meindl, P.: Supply Chain Management. Prentice-Hall (2001)

Chow, W.S., Madu, C.N., Kuei, C., Lu, M.H., Lin, C., Tseng, H.: Supply chain management in the US and Taiwan: an empirical study. Omega **36**(5), 565–579 (2008)

Clegg, N.: Making the metaverse: what it is, how it will be built, and why it matters (2022)

Corbett, C.J.: How sustainable is big data. Prod. Oper. Manag. **27**(9), 1685–1695 (2018)

Davenport, T.H.: Competing on analytics [WWW Document]. Harvard Business Review (2006). https://hbr.org/2006/01/competing-on-analytics. Accessed 2 Feb 2016

Davenport, T.H., Harris, J.G.: Competing on Analytics: The New Science of Winning. Harvard Business School Press (2007)

Dawe, R.L.: An investigation of the pace and determination of information technology use in the manufacturing materials logistics system. J. Bus. Logist. **15**(1), 229–258 (1994)

Dezi, L., Santoro, G., Gabteni, H., Pellicelli, A.C.: The role of big data in shaping ambidextrous business process management: Case studies from the service industry. Bus. Process. Manag. J. **24**(5), 1163–1175 (2018). https://doi.org/10.1108/BPMJ-07-2017-0215

Dierickx, I., Cool, K.: Asset stock accumulation and sustainability of competitive advantage. Manage. Sci. **35**(12), 1504–1511 (1989). https://doi.org/10.1287/mnsc.35.12.1504

Dolgui, A., Ivanov, D.: Metaverse supply chain and operations management. Int. J. Prod. Res. (2023). https://doi.org/10.1080/00207543.2023.2240900

Dubey, R., et al.: Can big data and predictive analytics improve social and environmental sustainability? Technol. Forecast. Soc. Chang. **144**, 534–545 (2019). https://doi.org/10.1016/j.techfore.2017.06.020

Dwivedi, Y.K., et al.: Metaverse beyond the hype: multidisciplinary perspectives on emerging challenges, opportunities, and agenda for research, practice and policy. Int. J. Inf. Manage. **66**, 102542 (2022). https://doi.org/10.1016/j.ijinfomgt.2022.102542

Dwivedi, Y., et al.: How metaverse will change the future of marketing: implications for research and practice. Psychol. Mark. (2023). https://doi.org/10.1002/mar.21767

Fadhil, S.S., Ismail, R., Alnoor, A.: The influence of soft skills on employability: a case study on technology industry sector in Malaysia. Interdiscip. J. Inf. Knowl. Manag. **16**, 255 (2021)

Fantini, P., Pinzone, M., Taisch, M.: Placing the operator at the centre of Industry 4.0 design: Modelling and assessing human activities within cyber-physical systems. Comp. Indust. Eng. (2018)

Ferraris, A., Mazzoleni, A., Devalle, A., Couturier, J.: Big data analytics capabilities and knowledge management: impact on firm performance. Manag. Decis. **57**(8), 1923–1936 (2019). https://doi.org/10.1108/MD-07-2018-0825

Gandomi, A., Haider, M.: Beyond the hype: big data concepts, methods, and analytics. Int. J. Inf. Manage. **35**(2), 137–144 (2015). https://doi.org/10.1016/j.ijinfomgt.2014.10.007

Gartner. Gartner Says Worldwide IT Spending on Pace to Grow 3.2 Percent in 2014 (2014)

Gatea, A.A., Marina, V.: Higher education funding in Iraq in terms of the experience of particular developed countries. Int. J. Adv. Stud. **6**(1), 8–17 (2016)

George, G., Haas, M.R., Pentland, A.: Big data and management. Acad. Manag. J. **27**, 321–326 (2014)

Gharajeh, M.S.: Biological big data analytics. In: Advances in Computers, vol. 109, pp. 321–355. Elsevier, Amsterdam (2018). ISBN 9780128137864

Giddings, B., Hopwood, B., O'Brien, G.: Environment, economy, and society: fitting them together into sustainable development. Sustain. Dev. **10**, 187–196 (2002)

Gobble, M.M.: Big data: the next big thing in innovation. Res. Technol. Manag. **56**, 64–66 (2013)

Gobbo, S.C.D.O., Fusco, J.P.A., Junior, J.A.G.: An analysis of embeddedness in the value creation in interorganizational networks: an illustrative example in Brazil. Int. J. Adv. Oper. Manag. **6**(2), 178–198 (2014)

Goes, P.B.: Big data and IS research. MIS Q. **38**, 3–8 (2014)

Govindan, K., Azevedo, S.G., Carvalho, H., Cruz-Machado, V.: Impact of supply chain management practices on sustainability. J. Clean. Prod. **85**, 212–225 (2014). https://doi.org/10.1016/j.jclepro.2014.05.068

Grant, R.M.: The resource-based theory of competitive advantage: implications for strategy formulation. Calif. Manage. Rev. **33**(3), 114–135 (1991). https://doi.org/10.2307/41166664

Grover, V., Chiang, R.H.L., Liang, T., Zhang, D.: Creating strategic business value from big data analytics: a research framework. J. Manag. Inf. Syst. **35**(2), 388–423 (2018). https://doi.org/10.1080/07421222.2018.1451951

Hamid, R.A., et al.: How smart is e-tourism? A systematic review of smart tourism recommendation system applying data management. Comp. Sci. Rev. **39**, 100337 (2021)

Hart, S.L.: A natural-resource-based view of the firm. Acad. Manag. Rev. **20**(4), 986–1014 (1995). https://doi.org/10.5465/amr.1995.9512280033

Hart, S.L., Dowell, G.: Invited editorial: a natural-resource-based view of the firm: fifteen years after. J. Manag. **37**(5), 1464–1479 (2011). https://doi.org/10.1177/0149206310390219

Hart, S.L., Milstein, M.B.: Creating sustainable value. Acad. Manag. Perspect. **17**(2), 56–67 (2003). https://doi.org/10.5465/ame.2003.10025194

Hitt, M.A., Xu, K., Carnes, C.M.: Resource based theory in operations management research. J. Oper. Manag. **41**, 77–94 (2016). https://doi.org/10.1016/j.jom.2015.11.002

Hoskisson, R.E., Wan, W.P., Yiu, D., Hitt, M.A.: Theory and research in strategic management: swings of a pendulum. J. Manag. **25**(3), 417–456 (1999). https://doi.org/10.1016/S0149-2063(99)00008-2

Huo, B., Gu, M., Wang, Z.: Green or lean? A supply chain approach to sustainable performance. J. Clean. Prod. **216**, 152–166 (2019). https://doi.org/10.1016/j.jclepro.2019.01.141

Ibrahim, Y.M., Hami, N., Othman, S.N.: Integrating sustainable maintenance into sustainable manufacturing practices and its relationship with sustainability performance: a conceptual framework. Int. J. Energy Econ. Policy **9**(4), 30–39 (2019). https://doi.org/10.32479/ijeep.7709

Ivanov, D., Dolgui, A.: A digital supply chain twin for managing the disruption risks and resilience in the era of Industry 4.0. Prod. Plan. Control (2020)

Junaid, M., Zhang, Q., Syed, M.W.: Effects of sustainable supply chain integration on green innovation and firm performance. Sustain. Prod. Consump. **30**, 145–157 (2022)

Kache, F., Seuring, S.: Challenges and opportunities of digital information at the intersection of Big Data Analytics and supply chain management. Int. J. Oper. Prod. Manag. **37**(1), 10–36 (2017)

Kamble, S.S., Gunasekaran, A.: Big data-driven supply chain performance measurement system: a review and framework for implementation. Int. J. Prod. Res. **58**(1), 65–86 (2020)

Kamble, S.S., Gunasekaran, A., Gawankar, S.A.: Achieving sustainable performance in a data-driven agriculture supply chain: a review for research and applications. Int. J. Prod. Econ. **219**, 179–194 (2020)

Karim, S., Mitchell, W.: Path-dependent and path-breaking change: reconfiguring business resources following acquisitions in the U.S. medical sector, 1978–1995. Strateg. Manag. J. **21**(10–11), 1061–1081 (2000). https://doi.org/10.1002/1097-0266(200010/11)21:10/11%3c1 061::AID-SMJ116%3e3.0.CO;2-G

Kauffman, R.J., Srivastava, J., Vayghan, J.: Business and data analytics: new innovations for the management of e-commerce. Electron. Commer. Res. Appl. **11**, 85–88 (2012)

Khan, S.A.R., Kamble, S.S., Zkik, K., Belhadi, A., Touriki, F.E.: Evaluating barriers and solutions for social sustainability adoption in multi-tier supply chains. Int. J. Prod. Res. (2021). https://doi.org/10.1080/00207543.2021.1876271

Khaw, K.W., et al.: Modelling and evaluating trust in mobile commerce: a hybrid three stage Fuzzy Delphi, structural equation modeling, and neural network approach. Int. J. Hum. Comput. Interact. **38**(16), 1529–1545 (2022)

Kiron, D., Prentice, P.K., Ferguson, R.B.: The analytics mandate. MIT Sloan Manag. Rev. **55**, 1–25 (2014)

Kumar, V., Chibuzo, E.N., Garza-Reyes, J.A., Kumari, A., Rocha-Lona, L., Lopez-Torres, G.C.: The impact of supply chain integration on performance: evidence from the UK food sector. Procedia Manufact. **11**, 814–821 (2017)

Lee, J.G., Kang, M.: Geospatial big data: challenges and opportunities. Big Data Research **2**, 74–81 (2015)

Lehrer, C., Wieneke, A., Brocke, J., Jung, R., et al.: How big data analytics enables service innovation: materiality, affordance, and the individualization of service. J. Manag. Inf. Syst. **35**(2), 424–460 (2018). https://doi.org/10.1080/07421222.2018.1451953

Lenny Koh, S.C., Demirbag, M., Bayraktar, E., Tatoglu, E., Zaim, S.: The impact of supply chain management practices on performance of SMEs. Indust. Manag. Data Syst. **107**(1), 103–124 (2007). https://doi.org/10.1108/02635570710719089

Li, S., Rao, S.S., Ragu-Nathan, T.S., Ragu-Nathan, B.: Development and validation of a measurement instrument for studying supply chain management practices. J. Oper. Manag. **23**(6), 618–641 (2005). https://doi.org/10.1016/j.jom.2005.01.002

Liu, Y., Zhu, Q., Seuring, S.: Linking capabilities to green operations strategies: the moderating role of corporate environmental proactivity. Int. J. Prod. Econ. **187**, 182–195 (2017)

Lopes de Sousa Jabbour, A.B., Gomes Alves Filho, A., Backx Noronha Viana, A., José Chiappetta Jabbour, C.: Measuring supply chain management practices. Measur. Bus. Excell. **15**(2), 18–31 (2011). https://doi.org/10.1108/13683041111131592

Lunden, I.: Forrester: $2.1 Trillion Will Go Into IT Spend in 2013; Apps and the U.S. Lead the Charge (2013)

Lundmark, P.: The real future of the metaverse is not for consumers (2022). https://www.ft.com/content/af0c9de8-d36e-485b-9db5-5ee1e57716cb

Mangla, S.K., Kusi-Sarpong, S., Luthra, S., Bai, C., Jakhar, S.K., Khan, S.A.: Operational excellence for improving sustainable supply chain performance. Resour. Conserv. Recycling **162** (2020). https://doi.org/10.1016/j.resconrec.2020.105025

Manhal, M., Al-khalidi, A., Hamad, Z.: Strategic network: managerial myopia point of view. Manag. Sci. Lett. **13**(3), 211–218 (2023)

Martens, D., Provost, F., Clark, J.: Mining massive fine-grained behavior data to improve predictive analytics. MIS Q., **40**(4), 869–888 (2016). https://doi.org/10.25300/MISQ/2016/40.4.04

McAfee, A., Brynjolfsson, E.: Big data: the management revolution. Harv. Bus. Rev. **60–66**(68), 128 (2012)

Menguc, B., Ozanne, L.K.: Challenges of the "green imperative": a natural resource-based approach to the environmental orientation–business performance relationship. J. Bus. Res. **58**(4), 430–438 (2005). https://doi.org/10.1016/j.jbusres.2003.09.002

Minelli, M., Chambers, M., Dhiraj, A.: Big Data, Big Analytics: Emerging Business Intelligence and Analytic Trends for Today's Businesses. Wiley (2013)

Moktadir, M.A., Ali, S.M., Paul, S.K., Shukla, N.: Barriers to big data analytics in manufacturing supply chains: a case study from Bangladesh. Comp. Indust. Eng. **128**, 1063–1075 (2019). https://doi.org/10.1016/j.cie.2018.04.013

Müller, O., Fay, M., Brocke, J.: The effect of big data and analytics on firm performance: an econometric analysis considering industry characteristics. J. Manag. Inf. Syst. **35**(2), 488–509 (2018). https://doi.org/10.1080/07421222.2018.1451955

Narasimhan, R.: Strategic supply management: a total quality management imperative. Adv. Manag. Organ. Quality **2**, 39–86 (1997)

Nidumolu, R., Prahalad, C.K., Rangaswami, M.R.: Why sustainability is now the key driver of innovation. Harv. Bus. Rev. **87**(9), 56–64 (2009)

Nilsson, F., Göransson, M.: Critical factors for the realization of sustainable supply chain innovations—model development based on a systematic literature review. J. Clean. Prod. **296**, 126471 (2021)

Olszak, C.M.: Towards an understanding business intelligence. A dynamic capability-based framework for Business Intelligence. In: Proceedings of the 2014 Federated Conference on Computer Science and Information Systems (FedCSIS 2014), Warsaw, 7–10 September 2014 (2014)

Ou, C.S., Liu, F.C., Hung, Y.C., Yen, D.C.: A structural model of supply chain management on firm performance. Int. J. Oper. Prod. Manag. **30**(5), 526–545 (2010)

Penrose, E.T.: The Theory of the Growth of the Firm. Oxford University Press (2009)

Queiroz, M.M., Fosso Wamba, S., Pereira, S.C.F., Chiappetta Jabbour, C.J.: The metaverse as a breakthrough for operations and supply chain management: implications and call for action. Int. J. Oper. Prod. Manag. **43**(10), 1539–1553 (2023). https://doi.org/10.1108/IJOPM-01-2023-0006

Retrieved from https://nickclegg.medium.com/making-the-metaverse-what-it-is-how-it-will-be-built-and-why-it-matters-3710f7570b04

Sarkis, J., Gonzalez-Torre, P., Adenso-Diaz, B.: Stakeholder pressure and the adoption of environmental practices: the mediating effect of training. J. Oper. Manag. **28**(2), 163–176 (2010). https://doi.org/10.1016/j.jom.2009.10.001

Schoenherr, T., Speier-Pero, C.: Data science, predictive analytics, and big data in supply chain management: current state and future potential. J. Bus. Logist. **36**(1), 120–132 (2015)

Schroeck, M., Shockley, R., Smart, J., Romero-Morales, D., Tufano, P.P.: Analytics: The Real-World Use of Big Data. IBM Institute for Business Value, New York (2012)

Seuring, S.A.: Assessing the rigor of case study research in supply chain management. Supply Chain Manag. Int. J. **13**(2), 128–137 (2008)

Shahbaz, M., Gao, C., Zhai, L., Shahzad, F., Luqman, A., Zahid, R.: Impact of big data analytics on sales performance in pharmaceutical organizations: the role of customer relationship management capabilities. PLoS ONE **16**, e0250229 (2021)

Shamim, S., Zeng, J., Khan, Z., Zia, N.U.: Big data analytics capability and decision-making performance in emerging market firms: the role of contractual and relational governance mechanisms. Technol. Forecast. Soc. Chang. **161**, 120315 (2020)

Sharma, K., Giannakos, M.: Multimodal data capabilities for learning: what can multimodal data tell us about learning? Br. J. Edu. Technol. **51**(5), 1450–1484 (2020)

Shukla, A., Deshmukh, S., Kanda, A.: Environmentally responsive supply chains: learnings from Indian auto sector. J. Adv. Manag. Res. **6**(2), 154–171 (2009)

Sroufe, R.: Effects of environmental management systems on environmental management practices and operations. Prod. Oper. Manag. **12**(3), 416–431 (2003). https://doi.org/10.1111/j.1937-5956.2003.tb00212.x

Stevens, G.: Integrating the supply chains. Int. J. Phys. Distrib. Mater. Manag. **8**(8), 3–8 (1989)

Strawn, G.O.: Scientific research: how many paradigms? Educ. Rev. **47**, 26 (2012)

Tseng, M.-L., Wu, K.-J., Lim, M.K., Wong, W.-P.: Data-driven sustainable supply chain management performance: a hierarchical structure assessment under uncertainties. J. Clean. Prod. **227**, 760–771 (2019)

Vachon, S., Mao, Z.: Linking supply chain strength to sustainable development: a country-level analysis. J. Clean. Prod. **16**, 1552–1560 (2008)

Venkatesh, V.G., Kang, K., Wang, B., Zhong, R.Y., Zhang, A.: System architecture for blockchain based transparency of supply chain social sustainability. Robot. Comput. Integrat. Manufact. **63** (2020). https://doi.org/10.1016/j.rcim.2019.101896

Waller, M.A., Fawcett, S.E.: Data science, predictive analytics, and big data: a revolution that will transform supply chain design and management. J. Bus. Logist. **34**(2), 77–84 (2013)

Wang, Y., Hajli, N.: Exploring the path to big data analytics success in healthcare. J. Bus. Res. **70**, 287–299 (2017)

Wernerfelt, B.: A resource-based view of the firm. Strateg. Manag. J. **5**(2), 171–180 (1984). https://doi.org/10.1002/smj.4250050207

Winter, S.G.: The satisficing principle in capability learning. Strateg. Manag. J. **21**(10–11), 981–996 (2000). https://doi.org/10.1002/1097-0266(200010/11)21:10/11%3c981::AID-SMJ125%3e3.0.CO;2-4

Wixom, B.H., Yen, B., Relich, M.: Maximizing value from business analytics. MIS Q. Exec. **12**, 111–123 (2013)

Wong, C.Y., Arlbjorn, J.S., Johansen, J.: Supply chain management practices in the toy supply chain. Supply Chain Manag. Int. J. **10**(5), 367–378 (2005)

Wook Kim, S.: Effects of supply chain management practices, integration and competition capability on performance. Supply Chain Manag. Int. J. **11**(3), 241–248 (2006). https://doi.org/10.1108/13598540610662149

Wu, S.J., Melnyk, S.A., Calantone, R.J.: Assessing the core resources in the environmental management system from the resource perspective and the contingency perspective. IEEE Trans. Eng. Manage. **55**(2), 304–315 (2008). https://doi.org/10.1109/TEM.2008.919727

Zhong, R.Y., Newman, S.T., Huang, G.Q., Lan, S.: Big Data for supply chain management in the service and manufacturing sectors: challenges, opportunities, and future perspectives. Comp. Indust. Eng. **101**, 572–591 (2016)

Zhu, C., Du, J., Shahzad, F., Wattoo, M.U.: Environment sustainability is a corporate social responsibility: measuring the nexus between sustainable supply chain management, big data analytics capabilities, and organizational performance. Sustainability **14**, 3379 (2022)

Using Artificial Intelligence and Metaverse Techniques to Reduce Earning Management

Yahia Ali Kadhim(✉) and Safaa Ahmed Mohammed Al Ani

Department of Accounting, University of Baghdad, Baghdad, Iraq
Yahia0071985@gmail.com,
Prof.drsafaa_alani@coadec.uobaghdad.edu.iq

Abstract. This study aims to demonstrate the role of artificial intelligence and metaverse techniques, mainly logistical Regression, in reducing earnings management in Iraqi private banks. Synthetic intelligence approaches have shown the capability to detect irregularities in financial statements and mitigate the practice of earnings management. In contrast, many privately owned banks in Iraq historically relied on manual processes involving pen and paper for recording and posting financial information in their accounting records. However, the banking sector in Iraq has undergone technological advancements, leading to the Automation of most banking operations. Conventional audit techniques have become outdated due to factors such as the accuracy of data, cost savings, and the pace of business completion. Therefore, relying on auditing a large volume of financial data is insufficient. The Metaverse is a novel technological advancement seeking to fundamentally transform corporate operations and interpersonal interactions. Metaverse has implications for auditing and accounting practices, particularly concerning a company's operational and financial aspects. Economic units have begun to switch from traditional methods of registration and posting to using software for financial operations to limit earnings management. Therefore, this research proposes applying one of the Data Mining techniques, namely the logistical regression technique, to reduce earning management in a sample of Iraqi private banks, including (11) banks. Accounting ratios were employed, followed by Logistic Regression, to achieve earnings management within the proportions.

Keywords: Artificial Intelligence · Earnings Management · Metaverse · Logistic Regression

1 Introduction

Earnings management comprises a broad spectrum from conservatism in accounting-to-accounting deceit. Managers employ various tools and techniques to control earnings for diverse reasons and motivations (Mahmoudi et al., 2017). According to Watts and Zimmerman (1986), the management of profits by the administration serves several purposes, such as incentivizing bonus plans, complying with obligations related to debt, mitigating political expenses, and pursuing capital goals (Arkan, 2015). The management of earnings is assisted by the presence of motives and incentives that drive the administration to engage in such behavior. Management can employ various strategies, including altering the Generally Accepted Accounting Principles (GAAP) procedure, using the Big Bath approach, and making Big Bets for the Future to manage earnings effectively (Sulaiman et al., 2014).

Therefore, management of its earnings has severe consequences for the economic unit, as management of profits can reduce the credibility of accounting figures and thereby harm the monetary unit's reputation (Alves and Pereira, 2017). The audit assures the quality and credibility of the economic unit's financial information, as the auditors play two valuable roles for the participants of the capital market. The significance of data as well as the function of confirmation, as the auditors provide independent verification of the financial statements prepared by the economic unit. The auditor's quality contributes to the credibility of the financial information and thus limits the possibility of fraud. Earnings management is because high-quality auditors reduce earnings management (Alves, 2013). Accounting will be required to safeguard precious and restricted funds and ensure optimal utilization. In this aspect, some argue that the realm of Metaverse has its virtual economy, which uses the same scarcity concept as the real-world economy, implying that Metaverse needs the accounting profession. Others argue that the virtual world's nature makes it unsuited for scarcity-based economics and that replicating the financial system, in reality, is not the best route ahead in the world of Metaverse. For instance, computerized accounting systems and financial in general will modify how they achieve their aims in the Metaverse world, which will be reflected in the operations of both these programs: evaluation and accounting disclosure (Al-Gnbri, 2022; Al-Abrrow et al., 2021). Numerous artificial intelligence techniques can be used to reduce earning management, such as data mining to obtain more information by analyzing and extracting data and addressing the weakness of mathematical modeling by removing helpful information from massive data sets. The logistic Regression technique can also assist auditors in discovering and extracting concealed data patterns. The emergence of this technology has been facilitated by the advancement and accessibility of contemporary risk-based auditing and computer-assisted auditing methods (CAATs). Additionally, utilizing business information systems and various technological tools further supports implementing these technologies (Zaarour and Dbouk, 2017; Albahri et al., 2021). No studies have examined how artificial Intelligence, especially logistic Regression, can reduce earnings management in Iraqi private banks. Consequently, this research aims to explore the methods of using logistic regression techniques to help auditors detect earning management in Iraqi private banks. In this study, we emphasize the use of Artificial Intelligence and its techniques in Iraqi banks because auditing financial reports relies on traditional methods and does not use artificial Intelligence, as well as auditing processes that require more time and effort. Hence, this study makes the following contribution:

1. Introducing modern technology into auditing operations, as current technologies have not previously been used in auditing activities.
2. The application of artificial Intelligence and its methodologies in auditing bank financial reports would assist in minimizing effort and time and thus examine more samples, resulting in greater accuracy and efficiency in audit outcomes.
3. Because artificial Intelligence and its approaches have not been applied in auditing bank financial reports, this research will employ one of the synthetic intelligence

strategies. Consequently, this study will comprise an introduction, a literature review, a Hypothesis Development, a Methodology, results, a discussion, and a conclusion.

2 Literature Review

2.1 Earnings Management Concept

The trade-off between the relevance and dependability of reported accounting data is crucial regarding earning management. Focusing on materiality would emphasize the present value of expected future cash flows. Management would only disclose realized cash flows if concerned only with credibility. Relevance and reporting reliability are not mutually exclusive. However, as relevance increases, accounting data reliability tends to decrease. The issue is that a comprehensive report necessitates subjective management judgments. Because different estimations are difficult to verify, they are susceptible to manipulation by management. Despite the challenges of accurate reporting, the reported accounting data must be relevant and trustworthy. Therefore, generally recognized accounting principles allow for some reporting flexibility. The internal auditing process has emphasized firms' concern to leverage it to create additional jobs and employ new methodologies. They also utilize it as a consultant within the organization who acts as an advisor for management, as managers frequently face essential times such as earning management (Yousif and Mohamed, 2022; AL-Fatlawey et al., 2021). In general, earning management can be regarded as taking advantage of the reporting flexibility provided by generally accepted accounting principles (Hoglund, 2012; Alnoor et al., 2022).

The relationship between internal audit and company governance significantly impacts various economic activities. This association's anticipated meanings and repercussions have undergone significant changes in recent years (Yousif and Mohammed, 2022; Alsalem et al., 2022). The accounting profession emphasizes measuring and disseminating financial information to creditors, financiers, and regulators—participants in decisions about the economic entity. Therefore, external accounting should provide investors, creditors, regulators, customers, suppliers, and employees with beneficial information regarding their future investment and tax decisions. As a result of the managers' knowledge of the current internal state of the economic unit and the working conditions, the financial situation and performance of the monetary unit are portrayed accurately and fairly. Relevant and dependable accounting information is necessary for its use in decision-making. Due to the information asymmetry between managers and external consumers of accounting information, managers can use discretion when compiling and reporting accounting information for their benefit. Earnings management refers to control in gathering and reporting accounting information (Sanusi and Ghazali, 2015; Atiyah, 2020). So, Earnings manipulation refers to the intentional actions undertaken by the administration of economic entities to augment or diminish reported earnings in the financial statements of the monetary unit to deceive external users of financial statements through income manipulation and settlement and manipulation of revenues, expenses, gains, and losses to reduce earning fluctuations by manipulating financial information (Praptapa et al., 2021). Therefore, earnings profit refers to the strategic utilization of diverse accounting procedures to enhance the presentation of financial statements. Others believe that the most popular methods for earning management include the following (Tilling et al., 2012; Atiyah and Zaidan, 2022).

1. Accounting policy choice: the decision between fixed and accelerated depreciation methods, first-in-first-out or weighted average inventory valuation.
2. Accrual accounting: Rather than reporting irregular changes in revenue and earnings annually, managers prefer to achieve consistent growth in revenue and profits. Consequently, managers will be incentivized to use accrual accounting to manage earnings.
3. Smoothing of income: The objective is to reduce fluctuations in income from one year to the next by relocating earnings from years in which the economic unit achieves high gains to years in which the economic team earns reduced revenues. The economic unity's leadership employed revenue smoothing as the guiding concept for the identification of income and expenditure. This approach is also utilized in reserves management and the categorizing of usual and unusual elements in the income report (Al-Taie et al., 2017; Fadhil et al., 2021).

2.2 The Difference Between Earnings Management and Fraud

By Generally Accepted Accounting Principles (GAAP), earning management occurs through strict or conservative accounting at the end of a fiscal year, such as underestimating or overestimating specific provisions. It may also result from aggressive or conservative economic decisions made by managers during the fiscal year to influence cash flows, such as accelerating or delaying sales. Consequently, earnings management can either increase or decrease reported earnings. Earnings management can also be detrimental if it contributes to minimizing the economic unit's value, but it can be beneficial if it leads to future information about the monetary unit. Nonetheless, earnings management does not violate any accounting principles. On the other hand, fraud transpires when there is a violation of widely recognized accounting norms, such as the deliberate omission of obligatory provisions or the inclusion of fabricated or cancelled sales. Managers may perpetrate fraud to increase or decrease reported earnings during or after the fiscal year. Fraud typically follows aggressive earnings management, which is significantly more aggressive than earnings management (El Diri, 2018; Gatea and Marina, 2016).

Profits are the "final sum" to measure the economic unit's performance. It reveals to consumers of financial statements the extent to which the economic entity participates in activities that increase its value. The financial press provides numerous examples of profits and explains why profits deviate from expectations. Analysts and financial managers expect profits, as they serve as indicators of the economic unit's value. An upsurge in profits signifies an augmentation in value, whereas a decline in profits signifies a reduction in value. Shareholders utilize earnings to evaluate managerial performance, forecast future cash flows, and assess risks. Notably, profits exhibit a stronger correlation with stock prices than other factors. Consequently, both analysts and financial managers believe in the attainment of profits. As lenders use profits in debt covenants to reduce lending risks and monitor performance from cash flows or sales, customers may use profits to assess whether products and services will be provided. In contrast, employees use profits to determine the economic unit's prospects and the degree of job security they can maintain. Earnings management impacts the integrity of accounting information and earnings quality (Tilling et al., 2012; Hamid et al., 2021). Financial effectiveness is a key notion that is crucial in all commercial organizations, especially in financial companies. It provides a comprehensive and extensive overview of the company's operations, both internally and externally.

Profitability is a fundamental and all-encompassing term that is crucial for all firms, regardless of their field or expertise, whether it is accounting or administration. Financial performance refers to the evaluation and assessment of an entity's level of financial success, taking into account the specific methods and principles used for analysis. The disagreement on the notion of corporate performance is thought to arise from variations in the principles and criteria used to analyze and evaluate performance) Khadim Mohammed and Mohammed, 2022; Khaw et al., 2022). The net profit statement is the most crucial financial statement, serving as an index of the company's value addition. It directly influences the value of its stock and is very influential for decision-makers. This is because it significantly affects the performance evaluations conducted by many users. Furthermore, the quality of earnings has garnered attention from various stakeholders due to its substantial influence on economic decision-making. Consequently, there is no established procedure to quantify it and no precise characterization of the excellence of accounting revenues (Hameed et al., 2019; Manhal et al., 2023).

There exist two main categories of earnings management. One method manager employ to manipulate their profits is through actual earnings management. The strategic direction of economic earnings is when managers make deliberate decisions about the operational operations of a business. Economic profit is predicated on well-established opportunity costs in economic theory. Managers are aware that they must tolerate a loss of future cash flow when employing economic earning management techniques to accomplish the required short-term ratios. Other authors concur that business administrators must be well-informed on all aspects of financial performance and the associated costs to enhance an enterprise's financial performance. Economic earning management is more difficult to detect because it entails investment and operational decision-making strategies that impact a company's cash flow. The second type of Earnings management is accrual-based, which translates to accounting earning management. Accounting earning management is based on accruals, designed to present the business's proper performance by documenting revenue and expenses in the period they were incurred (Strakova, 2020).

2.3 Artificial Intelligence

Artificial Intelligence is a strategy or technique for emulating human Intelligence in a computer, an automaton, or a product. Artificial Intelligence studies how the brain learns, makes decisions, and solves problems. Artificial Intelligence seeks to enhance computer capabilities associated with human knowledge, such as reasoning, learning, and problem-solving. This artificial Intelligence's characteristics are ethereal. Intelligence's components are reasoning, learning, problem-solving, perception, and linguistic Intelligence (Gursoy et al., 2022). Knowledge representation, planning, natural language processing, learning, deductive reasoning, manifestation, and the ability to move and manipulate objects are the objectives of Artificial Intelligence research. There are definitive goals within the discipline of general Intelligence. Numerous methods are available, including artificial Intelligence, statistical methods, and conventional coding (Koo et al., 2023). Artificial Intelligence extensively uses synthetic neuronal circuits and practices based on economics, statistics, and probability. The study of research and mathematical optimization. As it has for all other industries, artificial Intelligence substantially impacts the accounting industry. AI-enabled accounting systems will assist finance professionals and their companies to remain competitive and attract the next generation of employees and customers by saving time and money and providing insights. Research and development in machine learning have experienced rapid expansion in recent decades. Accounting, "robotics, natural language processing, expert systems, speech recognition, computer vision," Considerable advancements have been achieved, notwithstanding unresolved issues of notable magnitude (Buhalis et al., 2023). Artificial Intelligence's success is partly due to the development of novel system architectures that can utilize all the knowledge, including human expertise, in a given domain. Thus, these knowledge-based systems take into account human expertise to enhance their performance. Today's businesses embrace and implement new technologies to advance their operations, with accountability being one of their top priorities—artificial Intelligence yields positive outcomes such as increased and enhanced productivity, higher-quality precision, and reduced costs. Chat programs, Machine learning tools, Automation, and other Artificial Intelligence technologies play a crucial role in the accounting industry. By investing heavily in these technologies, accounting firms make them an integral part of their operations. Artificial intelligence applications have an impact on accounting professionals' daily tasks (Go and Kang, 2023).

Using Artificial Intelligence, accountants can increase their productivity, efficiency, and ability to acquire new clients. Artificial Intelligence can supplant humans in mundane and repetitive data extraction, structuring, organization, and configuration tasks. However, duplicate accounts and auditors utilizing Artificial Intelligence can perform various duties (Um et al., 2022). Artificial Intelligence is first taught what data to search for and how to organize it. Then, they investigate flaws and anomalies. Thus, artificial Intelligence can perform repetitive tasks that consume much time due to various data input and reconciliation processes, eliminate errors, and reduce liability. Accountants will be able to focus on more desirable responsibilities (Clifford and Janaki, 2021).

In 1956, at a symposium held at Dartmouth College in the United States, John McCarthy and other computer experts presented a proposal on the concept of "artificial intelligence" to the International Joint Conference on Artificial Intelligence to investigate suitable technologies (Cai and Luo, 2018). Artificial Intelligence is one of the disciplines of study that focuses on developing computer hardware and software that can act intelligently and perform duties and tasks more accurately and effectively than humans. Artificial Intelligence refers to developing and implementing computer systems configured to do activities that require human cognitive abilities. It comprises the ability to discriminate, comprehend relationships, and generate ideas (Amaka and Nnenna, 2020). There are many different technologies used in artificial Intelligence and other disciplines, such as (Gera et al., 2020).

1. Artificial Neural Networks (ANNs) are computer models designed to mimic the structure and functionality of the human organic central nervous system. ANNs comprise several linked processing components, or neurons, which work together to address specific problem-solving tasks.
2. Back Propagation Algorithm: This algorithm progressively seeks to reduce errors to a minimum, and training continues until mistakes are acceptable.
3. Self-Organizing Maps: An unsupervised neural network computational mapping technique that creates ordered non-linear models of high-dimensional data elements, typically one-dimensional or two-dimensional.
4. Fuzzy logic is a method for modelling imprecise models of reasoning, such as logical thinking for ambiguous and complex processes. This technique generates decisions based on approximate information and uncertainty, much like human thought.
5. Genetic Algorithm: It is an approach to machine learning based on the theory of evolutionary computation that simulates the process of natural selection and has a remarkable capacity to solve a problem (Kalogirou and Mellit, 2008).
6. Logistic Regression is a classification procedure commonly employed to estimate the likelihood of dependent variables that have been categorized. Logistic Regression involves utilizing a dependent variable characterized by binary values, expressed explicitly as 1 to represent success or 0 to denote failure (Payal et al., 2020).
7. Support Vector Machine: It is a classification approach commonly used in data mining and is considered most of the machine learning techniques that have become widely used (Ibrahim and Jaber, 2022).
8. Classification and regression tree: This methodology is non-parametric and includes the study of both category and mathematical variables (Yang and Wang, 2023). CART analysis is employed in classification and regression problems. Its main objective is to create models that utilize anticipated variables to forecast the amounts of the answer factor in a non-parametric manner. The analysis involves a binary growth process that is iteratively applied to an information set based on predetermined division standards (Khalil and Rada, 2023).

New technologies are significantly impacting operations across all industries. Furthermore, it can modify customers' expectations during their interactions with firms. Similarly, the same principle applies to the field of accounting. Artificial Intelligence can potentially enhance accountants' productivity and efficiency. By significantly decreasing the duration needed to fulfil responsibilities by 80–90%, accountants will have the capacity to allocate a more significant portion of their time towards providing advisory services to customers. Integrating artificial Intelligence into accounting procedures will improve the quality by mitigating mistakes. Integrating artificial intelligence technology into the operations of accounting companies enhances their attractiveness as employers and service providers to customers. As the integration of artificial Intelligence becomes prevalent among accounting companies, those that embrace this technology will have the capacity to offer data enabled by Automation, gaining a competitive advantage over those who resist its use. Robotic Process Automation (RPA) allows automated robots execute labor-intensive and repetitive tasks inside company operations, including the analysis and processing of documents. Intelligent Automation represents a more sophisticated iteration of the Automation of robotic processes. Innovative technology can replicate human interaction in many scenarios, shown by its ability to perceive the intended significance of a customer's message and adjust its behaviour based on prior data. Robotic Process Automation (RPA) and intelligent Automation are widely utilized in accounting, including many applications. This technology can efficiently process documents via machine translation and computer vision, surpassing previous speed levels.

Additionally, it often can promptly furnish up-to-date information on financial problems, enabling the generation of daily reports cost-effectively. This understanding allows organizations to proactively adjust their trajectory in response to data indicating unfavourable patterns. The use of robotic machine learning for automatic permission and document processing can potentially boost several internal accounting operations, including but not limited to buying, informing, order processing, expense reports, and payable and receivable accounts. Adding too many internal business regulations and municipal, state, and federal guidelines is imperative in accounting. AI-enabled solutions play a crucial role in facilitating audits and promoting conformity by enabling the meticulous tracking of documents by relevant rules, legislation, and indicators of potential non-compliance. Machine learning algorithms provide the capability to analyze extensive datasets efficiently, enabling the identification of possible instances of fraud or suspicious activities that human observers may have disregarded. These algorithms promptly flag such cases for subsequent examination (Madina, 2021).

2.4 The Application of Artificial Intelligence in the Field of Accounting

The categorization of accounting as a career with a substantial likelihood of Automation, a viewpoint occasionally substantiated by scholarly and practical empirical investigations, is undeniably a matter of apprehension. However, it should disregard these gloomy prophecies because the profession is anything but a path to disgrace. Artificial Intelligence ought to be regarded as the onset of a renaissance, showcasing its capacity to adapt to contemporary shifts in the corporate landscape and management demands. Intelligent systems can offer advantageous outcomes for accountants. Their skills enable them to resolve three major issues: support the decision-making process by providing better and cheaper data, provide deeper data analysis and new business insights, and prioritize more critical tasks after saving time with AI applications. As previously stated, the proper method for evaluating the prospect of an employment change is to examine the accounting work order's content. Undoubtedly, intelligent systems can and will replace specific duties; the question is when and to what extent. Recent advancements in artificial Intelligence and AI's ability to address real-world accounting issues render current research and implementation initiatives insufficient. The most recent findings from the Computer Science Department's 2017 report on computer science and the prospective trajectory of the accounting field. Bookkeeping is widely recognized as the most regular and time-intensive aspect of accounting and is undoubtedly suitable for robotics. The double input logic enables the accounting entries to be uniquely coded. Complex business transactions can be easily categorized, described, and recorded in accounting records. The entire process is automatable. The precision of accounting information and the velocity of registrations will be enhanced. Sales forecasting is another task where AI could potentially be useful. Forecasting is a complex undertaking due to the presence of knowledge asymmetry and associated dangers despite the existence of current models and techniques. The importance of precise sales forecasting cannot be overstated, as it directly impacts allocating resources within the operational budget and all related expenses. Using prediction models that rely on algorithms for machine learning can potentially improve the accuracy and reliability of predictive data. As a result, this may lead to enhancements in the planning and process of strategic management. Due to the possibility of intrinsic errors, accountants must give special attention to the dataset quality used for forecasting and planning. They must exercise reasonable care when submitting data to models (Todorova, 2018).

The accuracy of accounting can be influenced by various factors, including the standards of accounting, the economic and governmental systems, and the motivations of the financial statements. While it is acknowledged that transitioning to the International Financial Reporting Standards (IFRS) can affect the financial report, it remains the sole determinant of accounting quality. Conversely, applying IFRS may result in varying effects on accounting integrity across different nations due to differences in other influencing factors (Mohammed et al., 2020).

Examining job-task relocation necessitates an investigation of the pragmatic obstacles encountered by the accounting profession, particularly the acquisition of a novel skill set essential for executing tasks in a swiftly evolving landscape driven by digital technology and increasing data analysis capabilities. Integrating new intelligent technologies into business and accounting software has been driven by the fast transformation of the corporate environment resulting from the widespread use of AI applications. Consequently, many accountants have met these technologies without a comprehensive grasp of their underlying logic, capabilities, and full potential. The acquisition of technical proficiency in machine learning is highly valued, with the extent of knowledge required varying based on factors such as the size of the organization, its investment practices, and approach to fostering innovation. Notwithstanding these circumstances, accountants need to understand data quality's importance comprehensively. Machine learning involves identifying and using patterns derived from existing data points or instances, along with algorithms' continuous enhancement and adjustment over time. As previously indicated, instructing a computer using information sets requires meticulous deliberation on their precision. Many internal auditing measures may be employed to address the risks associated with bias and other inherent limitations in AI technologies (Todorova, 2018).

2.5 Artificial Intelligence in Auditing

The process of making decisions should be comprised of three essential iterative phases. The elements mentioned earlier encompass Intelligence (comprising data collection, goal setting, problem diagnosis, data validation, and problem structuring), design (surrounding data processing, goal setting, alternative generation, and risk or value assignment to alternatives), and selection (encompassing statistical analysis of other options, simulation of alternative outcomes, explanation of alternatives, selection among alternatives, and justification of choice). As a result, artificial Intelligence plays a crucial role within the decision-making domain since it is consistently being advanced and integrated into the technological and managerial aspects of contemporary enterprises and occupations, such as auditing. Consequently, artificial Intelligence and expert systems are advantageous and possibly essential for the current audit procedure. The primary objective of these systems is to enhance auditors' decision-making capabilities by mitigating the risks associated with biases and omissions that may arise in human decision-making procedures. Given the intricate nature and delicate nature of these evaluations, there is a prevailing belief that these systems should serve as inputs or assists in the auditor's final judgment of audit results.

Nevertheless, empirical research has observed that auditors occasionally depend excessively on the outcomes or results. Irrespective of the specific methods and methodologies the auditor utilizes in their analysis leads to a particular result or view, the auditor ultimately bears responsibility for their judgment. Like auditors who depend on the expertise of other professionals, such as appraisers and real estate attorneys, to gather audit evidence and form audit judgments, auditors employ artificial Intelligence technologies that function as hired "agents" assigned to specific tasks. The auditor is responsible for assuring these instruments' appropriateness, reliability, and effectiveness for their intended purpose. Moreover, using artificial Intelligence-based systems in decision-making processes has advantages and disadvantages. The auditor may bear responsibility for the inappropriate use of a contemporary decision aid, resulting in an entirely erroneous judgment and relying on an expert system to arrive at an improper conclusion (Omoteso, 2012). Intelligent systems can manage a large quantity of data, apply multiple methods to solve problems, and execute many audit jobs that people can only undertake in groups of several individuals (Mohammed and Abdullah, 2022).

2.6 Metaverse

The concept of the Metaverse was initially introduced more than three decades ago in Neal Stephenson's science fiction novel Snow Crash. The information technology sector has shown significant interest in the concept of Metaverse because of the quick advancements in technologies such as Blockchain, Web of Things (IoT), virtual reality/augmented reality, machine learning, and Cloud/Edge Technology. The online sandbox platform Roblox is credited as being the first firm to incorporate the word "metaverse" into its strategic framework. In doing so, they outlined many critical attributes of the Metaverse, including aspects such as identity, social connections, immersive experiences, minimal distractions, a civic environment, economic interactions, ubiquitous accessibility, and diverse content offerings. The social media platform Facebook has undergone a rebranding process, adopting the name Meta. This strategic move aims to facilitate the realization of the metaverse concept, whereby individuals may engage in novel forms of interaction, collaboration, studying, and socializing that were previously inconceivable (Yang, 2022). As a new social form, the Metaverse incorporates several modern technologies and possesses multi-technology features (Nesrine and Mohammed, 2023).

1. Immersive, realistic experience.
2. A comprehensive world structure.
3. Significant substantial economic value.
4. New rules and restrictions.
5. Significant uncertainty.

2.7 Artificial Intelligence and Metaverse

The scientific subject of artificial Intelligence originated with the fundamental assumption that all aspects of learning may potentially be accurately specified. Contemporary research in the field of Artificial Intelligence mostly centers on the domains of deep learning, machine learning, and reinforcement learning. These areas of study find application in several disciplines, including but not limited to image recognition, decision-making processes, and natural language processing (NLP). Individuals are inclined to pursue the actualization of the Metaverse through the progression of intelligent machinery inside the tangible realm. Machine learning serves as technical support inside the Metaverse, enabling systems to attain or match the degree of human knowledge. This development will significantly influence the Metaverse's operational efficiency and cognitive capabilities (Yang, 2022). When firms utilize virtual reality devices to generate financial records inside the global Metaverse, several aspects necessitate consideration. These characteristics have the potential to provide challenges to the accounting discipline soon (Al-Gnbri, 2022). The paramount significance of the Metaverse lies in its economic activities. Theoretically, the Metaverse should possess interoperability, enabling users to trade cross-platform virtual goods like clothing or automobiles. The subject matter under consideration pertains to digital assets, markets, and currency (Yang, 2022).

3 Hypothesis Development

Economic units can detect earnings management manipulation using artificial intelligence techniques such as machine learning and the Bayesian Naive Classifier. This research indicates that auditors should employ artificial Intelligence to detect earnings manipulation (Park and Kang, 2021). Other studies believe that techniques based on artificial Intelligence can aid in discovering the administration of corporate earnings and predicting them by utilizing neural network technology and the Bayesian Nave Classifier (Kao and Hsieh, 2016). Earning management in economic units can be detected through data mining techniques. Among these techniques are ANN, CHAID Decision Tree, and C5.0 Decision Tree, through which earnings management detection models are developed with more accuracy. The research revealed that the ANN/C5.0 model has the highest accuracy rate for detecting earning management. One of the primary challenges associated with forecasting revenue manipulation is the inherent imbalance within the available data. Supervised and unsupervised learning were employed to predict earnings management. The present study has made a significant contribution to the field of simulation-based selection, hence enhancing its applicability to several additional scenarios involving inaccurate information (Kumar et al., 2018).

As an earning management reducing mechanism, implementing information technology is crucial. The advent of the Industrial Revolution catalyzed the advancement and refinement of information technology, leading to the development of artificial intelligence systems that resemble human cognitive capabilities. The cognitive abilities of humans to store and analyze data for decision-making purposes are constrained. The organization is dedicated to profit management, which can have internal and outside ramifications. One of the apparent consequences is the erosion of confidence among investors and stakeholders in the company. The use and integration of technology for information inside firms represent a viable approach to mitigating the potential occurrence of managing earnings and using more efficient, integrated, and real-time interconnected accounting records. Using artificial Intelligence to detect and prevent earning management in the accounting process and financial statements is possible. The following types of artificial Intelligence have been implemented: A neural network refers to a computational model that aims to replicate the functioning of the human brain by simulating the activity of artificial neurons. This study focuses on the Multilayer Perception Algorithm, with particular emphasis on evaluating the efficacy of artificial neurons in predicting and categorizing improvements in systems. The integration of accounting technology is presently observed across all facets of accounting activities, giving rise to apprehensions over the future trajectory of the accounting field inside the organization.

Moreover, the most straightforward application of Artificial Intelligence is when computer programs can perform duties that humans typically perform manually. The advancement of artificial Intelligence is expected to be directly correlated with the level of intricacy involved in human labor, driving its further progress in the foreseeable future. One kind of machine learning is business intelligence and analytics technological advances, which gather data, analyze, and disseminate information to support decision-making. The incorporation of artificial Intelligence in auditing encompasses the utilization of Price Waterhouse Coopers' risk control workstation model and Deloitte's Visual Assurance, as observed from an auditing perspective (Yuniati and Mukti, 2021).

4 Methodology

The study utilized a representative group of eleven (11) Iraqi personal banks. The accounting records of each bank were carefully analyzed, and accounting ratios were calculated based on the data provided in the financial reports. Statistical methods were employed using the SPSS program to obtain specific ratios. Furthermore, logistic regression analysis was conducted on the financial ratios to explore their relationship with finance and the development of a linear formula for managing earnings.

5 Results

The statistical methodology employed aims to ascertain the presence of significant disparities among the means of two or more distinct and unrelated groups, and its application is limited to numerical information solely. The approach utilized in this study relies on the quantitative F test. Through the application of this analysis to the dataset, it was ascertained that the two variables, namely ACP/TL and CC/TD, possess moral attributes and may significantly and proficiently aid in discerning and revealing instances of earnings management inside the banking sector, as shown in Table 1.

It is a mathematical method to characterize the relationship between several independent variables and a single binary dependent variable. There is no earning management, as shown in Table 2.

The metric assesses the degree to which the estimated model can explain, namely the proportion of the overall variation in the dependent variable, income management that can be accounted for by the independent variables, as illustrated in Table 3.

Table 1. One Way ANOVA.

Description of the variable	Analysis of variance table						
Debt/equity	variables	sum of squares	degree of freedom	Mean squared error	F test	Sig	the decision
	D/E	0.463	1	0.463	0.199	0.666	No Sig
		20.894	9	2.322			
		21.357	10				
Debt/total assets	D/TA	0.011	1	0.011	0.401	0.543	No Sig
		0.243	9	0.027			
		0.254	10				
Profit before tax/total assets	PBT/TA	0	1	0	0.044	0.839	No Sig
		0.008	9	0.001			
		0.009	10				
Net profit/total assets	NP/TA	0	1	0	0.109	0.749	No Sig
		0.005	9	0.001			
		0.006	10				
Working capital/total assets	WC/TA	0.008	1	0.008	0.26	0.622	No Sig
		0.276	9	0.031			
		0.284	10				
Current assets/current liabilities	CA/CL	0.152	1	0.152	0.499	0.498	No Sig
		2.747	9	0.305			
		2.899	10				
Fixed assets/total assets	FA/TA	0.001	1	0.001	0.863	0.377	No Sig
		0.008	9	0.001			
		0.009	10				
cash/total assets	C/TA	0.042	1	0.042	1.317	0.281	No Sig
		0.285	9	0.032			
		0.327	10				
net profit/equity	NP/E	0.001	1	0.001	0.342	0.573	No Sig
		0.028	9	0.003			
		0.029	10				
accounts receivable/current asses	ACR/CA	0	1	0	0.511	0.493	No Sig
		0.004	9	0			
		0.004	10				
Revenue/total assets	R/TA	0.002	1	0.002	0.559	0.474	No Sig
		0.027	9	0.003			
		0.028	10				
net profit/revenue	NP/R	0.055	1	0.055	0.184	0.678	No Sig
		2.701	9	0.3			
		2.756	10				

(continued)

Table 1. (*continued*)

Description of the variable	Analysis of variance table						
Payables/total liabilities	ACP/TL	0.01	1	0.01	5.717	0.04	Sig
		0.015	9	0.002			
		0.025	10				
Cash Credit/total deposits	CC/TD	0.636	1	0.636	5.835	0.039	Sig
		0.981	9	0.109			
		1.617	10				
Net profit/paid capital	NP/PIC	0.005	1	0.005	0.426	0.53	No Sig
		0.097	9	0.011			
		0.102	10				
Investments/total deposits	Inv/TD	0.203	1	0.203	1.16	0.309	No Sig
		1.573	9	0.175			
		1.776	10				
Cash/equity credit	CC/E	2.289	1	2.289	2.056	0.185	No Sig
		10.019	9	1.113			
		12.307	10				

Table 2. Chi-square test results.

Omnibus Tests of Model Coefficients				
		Chi-square	df	Sig.
Step 1	Step	6.083	2	.048
	Block	6.083	2	.048
	Model	6.083	2	.048

Table 3. Displays the clarification coefficient's results.

Model Summary			
Step	−2 Log likelihood	Cox & Snell R Square	Nagelkerke R Square
1	6.808[a]	.425	.615

[a]Estimation terminated at iteration 7 because parameter estimates changed by less than .001

Table 1 consists of several financial ratios taken from the financial reports of several banks, and after using the One-Way ANOVA Table analysis, it was found that the two ratios (creditors/total liabilities and cash credit/total deposits) are the only two ratios that are significant because the degree of significance is (0.040, 0.039), respectively. It is less than 0.05, and these are the two ratios through which Earning management can be achieved. In this step, the regression curve's parameters are estimated, which includes the regression parameters for each variable, which is the percentage change in the dependent variable, earning management, if the independent variable changes by one unit.

The analysis results indicate the significance of the model's parameters by applying the Binary Logistic Regression model and the accompanying Chi-square test for the model's parameters. Table 2 is used to test the importance of the model parameters. The table shows that the model parameters are 95% significant, meaning that both percentages (creditors/total liabilities and cash credit/total deposits) are substantial because P-value = Sig. = 0.0480 < 0.05 and contribute to discovering Earning management.

Table 3 analysis results indicate that the two independent variables (creditors/total liabilities and cash credit/total deposits) explain 61.5% of the total variance of the earnings management variable, meaning that we can rely on 61.5% in the earnings management statement, which is a good percentage. In contrast, the remaining portion is 38.5%. It depends on other variables.

1. The estimated regression line equation will be in Table 4.

$$\ln\left(\frac{\hat{p}(x)}{1-\hat{p}(x)}\right) = -6.440 + 19.910ACP/TL + 4.815cc/tD$$

2. The researcher would like to point out that the linear regression curve equation can be used to detect financial fraud or not by applying it to any of the ratios mentioned above and any bank.
3. The ACP/TL variable contributes to a greater degree than the other variable; whenever the ACP/TL variable increases by one unit, the fraud detection rate increases by 19.91, and vice versa.
4. The variable cc/tD contributes to a lesser degree than the other variable; whenever the variable cc/tD increases by one unit, the fraud detection rate increases by 4.815, and vice versa.
5. The effect of the variable ACP/TL is significant according to the Wald test, as the p-value = sig. = 0.0401, which is a value less than 0.05. Thus, we reject the null hypothesis "no effect" and accept the alternative view "there is an effect." This effect is acceptable to a degree—95% confidence.
6. The effect of the variable cc/TD is significant according to the Wald test, as the p-value = sig. = 0.0251, which is a value less than 0.05. Thus, we reject the null hypothesis "no effect" and accept the alternative hypothesis "there is an effect." This effect is acceptable to a degree—95% confidence.

Table 4. The results of the model parameters

Variables in the Equation							
		B	S.E.	Wald	DF	Sig.	Exp (B)
Step 1[a]	ACP_TL	19.910	23.699	.706	1	.0401	443376393.284
	CC/TD	4.815	4.192	1.319	1	.0251	123.344
	Constant	−6.440	4.127	2.435	1	.0119	.002

[a]Variable(s) entered on step 1: ACP_TL, CC/TD

6 Discussion

The primary objective of this study was to investigate the impact of machine learning methodologies, namely logistic Regression, on mitigating the prevalence of earnings management among economic entities in Iraq. A hypothesis was formulated from an extensive review of existing literature. The study's results demonstrate a significant association between using Artificial Intelligence techniques and decreasing earnings management. This study aligns with prior research, as the utilization of artificial intelligence methods, namely logistic Regression, is an inevitable trajectory that will provide significant transformations and advancements to assist auditors in mitigating earnings management. This study aimed to investigate the effects of using artificial intelligence and metaverse techniques by auditors in the banking industry of Iraq, which is undergoing constant evolution. This adoption has directly influenced the operations and growth of the Iraqi banking sector. Based on the results of our study, it can be inferred that the utilization of the logistical regression approach has a significant impact on the reduction of earnings management. Consequently, it is recommended that auditors in Iraq incorporate this technique into their auditing practices. This work can be expanded by employing additional technologies, such as Blockchain technology, to reduce earnings management.

References

AL-Abrrow, H., Fayez, A.S., Abdullah, H., Khaw, K.W., Alnoor, A., Rexhepi, G.: Effect of open-mindedness and humble behavior on innovation: mediator role of learning. Int. J. Emerg. Market. (2021)

Albahri, A.S., et al.: Based on the multi-assessment model: towards a new context of combining the artificial neural network and structural equation modelling: a review. Chaos Solitons Fractals **153**, 111445 (2021)

AL-Fatlawey, M.H., Brias, A.K., Atiyah, A.G.: The role of strategic behavior in achievement the organizational excellence. Analytical research of the manager's views of Ur State Company at Thi-Qar Governorate. J. Administ. Econ. **10**(37) (2021)

AL-Gnbri, M.K.: Accounting and auditing in the metaverse world from a virtual reality perspective: a future research. J. Metaverse **2**(1), 29–41 (2022)

Alnoor, A., et al.: How positive and negative electronic word of mouth (eWOM) affects customers' intention to use social commerce? A dual-stage multi group-SEM and ANN analysis. Int. J. Hum. Comput. Interact. 1–30 (2022)

Alsalem, M.A., et al.: Rise of multiattribute decision-making in combating COVID-19: a systematic review of the state-of-the-art literature. Int. J. Intell. Syst. **37**(6), 3514–3624 (2022)

Al-Taie, B.F.K., Flayyih, H.H., Talab, H.R.: Measurement of income smoothing and its effect on accounting conservatism: an empirical study of listed companies in the Iraqi stock exchange. Int. J. Econ. Perspect. **11**(3), 710–719 (2017)

Alves, M.d.C.G., Pereira, A.: Earnings management and European Regulation 1606/2002: evidence from non-financial Portuguese companies listed in Euronext. E-mail address: mceu@ubi.pt (M.C.G. Alves), p. 5 (2017)

Alves, S.: The impact of audit committee existence and external audit on earnings management evidence from Portugal. J. Financ. Report. Account. **11**(2), 14 (2013)

Amaka, E., Modesta, N., Chukwuani, V.: Automation of accounting processes: impact of artificial intelligence. Int. J. Res. Innov. Soc. Sci. **IV**(VIII), 444 (2020)

Arkan, T.: The Effects of Earning Management Techniques, Net Income and Cash Flow on Stock Price. Scientific Papers of the University of Szczecin No. 855 Finance, Financial Markets, Insurance No. 74, vol. 2, p. 294 (2015)

Atiyah, A.G.: The effect of the dimensions of strategic change on organizational performance level. PalArch's J. Archaeol. Egypt/Egyptol. **17**(8), 1269–1282 (2020)

Atiyah, A.G., Zaidan, R.A.: Barriers to using social commerce. In: Artificial Neural Networks and Structural Equation Modeling: Marketing and Consumer Research Applications, pp. 115–130. Springer, Singapore (2022)

Buhalis, D., Leung, D., Lin, M.: Metaverse as a disruptive technology revolutionising tourism management and marketing. Tour. Manage. **97**, 104724 (2023)

Cai, Y., Meng, Q., Luo, J.: Analysis of the impact of artificial intelligence application on the development of accounting industry. Open J. Bus. Manag. 850–851 (2018)

Clifford, M., Janet, J.: A study on the scope of artificial Intelligence in accounting. Dogo Rangsang Res. J. UGC Care Group I J. **11**(05), 01 (2021)

El Diri, M.: Introduction to earnings management. Springer, Leeds (2018)

Fadhil, S.S., Ismail, R., Alnoor, A.: The influence of soft skills on employability: a case study on technology industry sector in Malaysia. Interdiscip. J. Inf. Knowl. Manag. **16**, 255 (2021)

Gatea, A.A., Marina, V.: Higher education funding in Iraq in terms of the experience of particular developed countries. Int. J. Adv. Stud. **6**(1), 8–17 (2016)

Gera, R., Srivastava, S., Tiwari, R.: Investigation of artificial intelligence techniques in finance and marketing. Int. Conf. Smart Sustain. Intell. Comput. Appl. Proc. Comput. Sci. **173**, 150–151 (2020)

Go, H., Kang, M.: Metaverse tourism for sustainable tourism development: tourism agenda 2030. Tourism Rev. **78**(2), 381–394 (2023)

Gursoy, D., Malodia, S., Dhir, A.: The metaverse in the hospitality and tourism industry: an overview of current trends and future research directions. J. Hospital. Market. Manag. **31**(5), 527–534 (2022)

Hameed, A.M., Al-taie, B.F.K., Al-Mashhadani, B.N.A.: The impact of IFRS 15 on earnings quality in businesses such as hotels: critical evidence from the Iraqi environment. Afr. J. Hospital. Tourism Leisure **8**(4), 1–11 (2019)

Hamid, R.A., et al.: How smart is e-tourism? A systematic review of smart tourism recommendation system applying data management. Comp. Sci. Rev. **39**, 100337 (2021)

Höglund, H.: Detecting earnings management with neural networks. Exp. Syst. Appl. (2012). https://doi.org/10.1016/j.eswa.2012.02.096

Ibrahim, M.H., Jaber, A.G.: The use of the regression tree and the support vector machine in the classification of the Iraqi Stock Exchange for the Period 2019–2020. J. Econ. Administ. Sci. **28**(132), 74–87 (2022)

Kalogirou, S., Mellit, A.A.: Artificial intelligence techniques for photovoltaic applications: a review. Prog. Energy Combust. Sci. **34**, 580 (2008)

Kao, Y., Hsieh, Y.-M.: An application of data mining techniques on earnings management detection. ICIC Exp. Lett. Part B: Appl. **7**(12), 2677 (2016)

Khadim Mohammed, A., Mohammed, S.A.: The impact of the Corona pandemic on the financial performance of companies listed on the Iraqi Stock Exchange. Resmilitaris **12**(2), 4897–4909 (2022)

Khalil, H.H., Rada, S.M.: An artificial intelligence algorithm to optimize the classification of the hepatitis type. J. Econ. Administ. Sci. **29**(135), 43–55 (2023)

Khaw, K.W., et al.: Modelling and evaluating trust in mobile commerce: a hybrid three stage Fuzzy Delphi, structural equation modeling, and neural network approach. Int. J. Hum. Comp. Interact. **38**(16), 1529–1545 (2022)

Koo, C., Kwon, J., Chung, N., Kim, J.: Metaverse tourism: conceptual framework and research propositions. Curr. Issue Tour. **26**(20), 3268–3274 (2023)

Kumar, U.D., Rahul, K., Seth, N.: Spotting earnings manipulation: using machine learning for financial fraud detection. In: Bramer, M., Petridis, M. (eds.) SGAI 2018. LNCS (LNAI), vol. 11311, pp. 343–356. Springer, Cham (2018). https://doi.org/10.1007/978-3-030-04191-5_29

Madina, E.: Artificial intelligence in accounting and auditing. Acad. J. Digit. Econ. Stabil. **1**(1) (2021)

Mahmoudi, A. Mahmoudi, S., Mahmoudi, S.: Prediction of earnings management using multilayer perceptron neural networks with two hidden layers in various industries. J. Entrep. Bus. Econ. 217 (2017). ISSN:2345-4695

Manhal, M., Al-khalidi, A., Hamad, Z.: Strategic network: managerial myopia point of view. Manag. Sci. Lett. **13**(3), 211–218 (2023)

Mohammed, B.H., Rasheed, H.S., Wahhab, R., Maseer, A.J.A.W.: The impact of mandatory IFRS adoption on accounting quality: Iraqi private banks. Int. J. Innov. Creat. Change **13**(5), 87–103 (2020)

Mohammed, E.J., Abdullah, S.H.: The quality of audit work under expert system. J. Econ. Administ. Sci. **28**(133), 187–199 (2022)

Nesrine, M., Mohammed, A.: Metaverse technique. Accounting practice in a virtual world. J. Res. Finance Account. **08**(1), 730–750 (2023)

Omoteso, K.: The application of artificial Intelligence in auditing: looking back to the future. Expert Syst. Appl. (2012). https://doi.org/10.1016/j.eswa.2012.01.098

Park, S., Kang, S.: Artificial Intelligence-Based Detection and Prediction of Corporate Earnings Management, p. 191. College of Business Administration, Ewha Womans University, Seoul (2021)

Payal, R., Saini, S., Kumar, Y.: Comparative analysis for fraud detection using logistic regression, random forest and support vector machine. Int. J. Res. Analyt. Rev. **7**(4), 728 (2020)

Praptapa, A., Pramuka, B.A., Dewi, R.S.: The influence of good corporate governance toward earnings management. Int. Sustain. Competitiv. Adv. 252 (2021)

Sanusi, Z.M., Shafie, N.A., Ghazali, A.W.: Earnings management: an analysis of opportunistic behaviour, monitoring mechanism and financial distress. In: 7th International Conference on Financial Criminology, 13–14 April 2015. Wadham College, Oxford (2015)

Strakova, L.: Earnings management in global background. SHS Web Conf. **2020**, 01032 (2020). https://doi.org/10.1051/shsconf/20207401032

Sulaiman, S., Danbatta, B.L., Abdul Rahman, R., Omar, N.: Management disclosure and earnings management practices in reducing the implication risk. Procedia Soc. Behav. Sci. **145**, 90–91 (2014)

Tilling, M., Ferlauto, K., McGowan, S., Stanton, P., Rankin, M.: Contemporary Issues in Accounting, 1st edn., p. 259. Wiley, Australia (2012)

Todorova, E.P.S.: How artificial Intelligence is challenging the accounting profession. J. Int. Sci. Publ. **12** (2018). ISSN:1314-7242

Um, T., Kim, H., Kim, H., Lee, J., Koo, C., Chung, N.: Travel Incheon as a metaverse: smart tourism cities development case in Korea. In: Stienmetz, J.L., Ferrer-Rosell, B., Massimo, D. (eds.) ENTER 2022, pp. 226–231. Springer, Cham (2022). https://doi.org/10.1007/978-3-030-94751-4_20

Yang, F.X., Wang, Y.: Rethinking metaverse tourism: a taxonomy and an agenda for future research. J. Hospital. Tourism Res. 10963480231163509 (2023)

Yang, Q., Zhao, Y., Huang, H., Xiong, Z., Kang, J., Zheng, Z.: Fusing blockchain and AI with metaverse: a survey. IEEE Open J. Comput. Soc. **3**, 122–136 (2022)

Yousif, N.S., Mohamed, S.A.: The role of internal audit in assessing the risks of management decisions regarding strategic operations acquisition. J. Econ. Administ. Sci. **28**(133), 172–186 (2022)

Yousif, N.S., Mohammed, S.A.: The role of internal auditing in governance of strategic operations and its reflection on management decisions. Resmilitaris **12**(2), 4910–4920 (2022)

Yuniati, T., Mukti, A.H.: How accounting artificial intelligence can prevent fraud (status and research opportunities). In: Conference on Management, Business, Innovation, Education and Social Science, vol. 1, no. 1 (2021)

Zaarour, I., Dbouk, B.: Towards a machine learning approach for earnings manipulation detection. Asian J. Bus. Account. **10**(2), 216–217 (2017)

The Metaverse and the Role of Accounting Culture: Reporting of Digital Assets According to International Standards

Ahmed Zuhair Jader[✉]

Accounting Techniques Department, Technical College of Management, Al-Furat Al-Awsat Technical University, 54003 Najaf, Iraq
ahmed.jader@atu.edu.iq

Abstract. The research paper aims to identify the accounting guide for what might be done when the eco-nomic unit conducts business deals of a new digital nature, setting up a guide for management that can be used for choosing the appropriate accounting policies for digital transactions. So, this guide is based on international bodies represented by the International Accounting Standards Board (IASB), these bodes and companies seek to put on the steps to accounting recognition initial of digital assets and, then reporting based on some international standards. For example, international accounting standard (IAS 2), international accounting standard (IAS 38) standard, or (IFRS 1), through the use of a digital bill from a company (BVNK) then using the steps by the accounting guide to making (accounting recognition and disclosure) of that digital data. The re-search concluded the possibility of creating a new accounting culture that keeps pace with digital developments by this guide and, also found out that the metaverse and its applications whether the blockchain, cloud computing and others, will have a role in reducing the costs, and saving the time to processing digital invoices, which helps to prepare and present financial statements for users in actual time, to meets their needs in getting high quality information and on-demand, become aware of the financial situation which enables them to make the correct economic decisions about continue or increase investment. This of course will be reflected in the continuity of the unit by keep their resources, as well as correcting management decisions regarding the appropriate use of resources.

Keywords: Metaverse · digital assets · virtual reality · augmented reality · financial Statements

1 Introduction

The world is always witnessing radical changes in the world of communication between the physical world and the digital world, or what is called the Fourth Industrial Revolution, and the appearance of technologies including artificial intelligence, cloud computing, and virtual reality, as well as robots, the Internet of Things, big data, blockchains, and other technologies that together form the meta-verse. The origin of the word meta-verse goes back to two words the first word (meta) which means beyond, and the other

M. Al-Emran et al. (Eds.): IMDC-IST 2024, LNNS 895, pp. 190–211, 2023.
https://doi.org/10.1007/978-3-031-51716-7_13

word (virus) means universe or world. (Al-Gnbri, 2022; Hamid et al., 2022). A number of researchers dealt with the subject of meta-verses, such as (Kalyvaki, 2023), about the Importance of the global trend towards digitization and benefiting from modern technologies in various fields, such as health, education, industry, tourism, economy, etc. It has become possible to communicate with the world from anywhere, without the need for the existence of an essence of matter in this world. Despite the modernity of the digital revolution, it has spread widely throughout the world. Through the meta-verse, imagine the size of the benefit that will be achieved for management, by obtaining timely information without waiting until the end of the fiscal year, and the ability to make investment decisions in light of several indicators in addition to accumulated experience, if they depend on the metaverse in practicing their activities (Al-Gnbri, 2022; Gibbs, 2021; Eneizan et al., 2019). Aharon et al. (2022) also argues that contrary to what is rumored that the metaverse is specialized in the field of games only, it goes beyond that to reach many sectors such as education, health, and others. Therefore, it is natural to achieve new revenues after adopting the meta-verse. As Aharon et al. (2023) mentioned the investment size might reach 82.02 billion dollars in 2023 which is expected to increase to 936.57 billion dollars in 2030. According to the company (Grand View Research), a consulting firm based in India and the United States, through their report in (April 2023; Al-Abrrow et al., 2021), it's announced the metaverse is a step towards an economic culture online (Grayscale, 2021).

The following chart shows investment in a number of professions, including gaming shares, which have a majority compared to other professions (Fig. 1.):

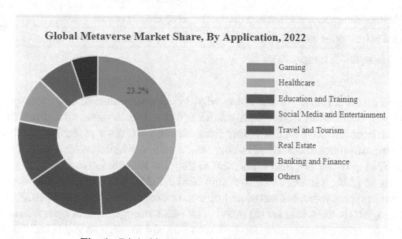

Global Metaverse Market Share, By Application, 2022

23.2%

- Gaming
- Healthcare
- Education and Training
- Social Media and Entertainment
- Travel and Tourism
- Real Estate
- Banking and Finance
- Others

Fig. 1. Digital investment in different professions.

Source: (Fortune Business Insights; date of visit: 30–07-2023; 01:20 A.M.).

Therefore, this research paper aims to shed light on what accounting might present, for transactions resulting from the two worlds (the virtual world with the real world), and also, on the difficulties that can be overcome through the flexibility that accounting characterizes in dealing with users' needs of information, then enabling the units to carry out their activities or develop them, through what they get of digital information, without delay in absorbing this change in the digital revolution, this is due to the wide open-ness witnessing companies, at various levels as a response of the meta-verse. Similarly (Baszucki, 2021; Khaw et al., 2022) discusses the future of business, which will change by holding virtual meetings without the need for physical attendance. (Table 1.) Among the types of metaverse technologies that can be displayed are:

Table 1. Metaverse technologies.

No	Metaverse technologies	Function
1	Artificial Intelligence (AI)	Create self-portraits and conversations
2	Virtual & Augmented Reality (VR/AR)	Virtual shopping instead of physical
3	Edge Computing (ED)	Data is processed before it reaches the cloud
4	5G Technology (5GT)	Data delivery between the source and the user in real time
5	Blockchain (B)	Avoid delays and decentralize grants
6	Internet of things (IT)	Seamlessly connecting the virtual world with the real one
7	Extended reality (ER)	Data formats in 3D formats
8	Brain-computer interfaces (BI)	Replacing traditional devices with what is digital and advanced
9	3D modeling and reconstruction (3DMR)	Preparing 3D models
10	Spatial and edge computing (SEC)	to respond to users

Source: (Shein, 2022; Dange, 2023)

Based on the table above and citing what Price Waterhouse & Coopers mentioned in its report, about what virtual reality and augmented reality are, virtual reality is the digital environment in which the user lives, through means and equipment used for virtual purposes that are described as augmented reality, which it is a support for virtual display by employing physical tools used to help the user understand the image in the digital world (Abbas et al., 2023; Bozanic et al., 2023). The virtual world appears in the form of graphs, a three-dimensional object, or others (Alsalem et al., 2022). On the other hand, jobs in the world are expected to increase through the emergence of software designers or programmers, for example (from 824,634 in 2019 to 23,360,639 in 2030). In addition, an improvement is expected. The global economy by $1.5 trillion by 2030. (PWC, 2019). Metaverse uses can be shown as in the following (Fig. 2.).

Source: Prepared by the researcher based on (Dwivedi et al., 2022).

The figure above shows the many uses of metaverse in many fields, including busi-ness and what can be used technically in this important sector. Thus, the provision of quality, credible, and timely information, helps the unit to develop its activities and the possibility of competition in light of the Fourth Industrial Revolution with other units within the same field (Haouam, 2020). Within the same context (Krishnamoorthy et al.,

Fig. 2. Metaverse classifications.

2020; Albahri et al., 2021) provide evidence of how the Internet of Things technology, for example, as one of the metaverse applications, collects a huge amount of data and then converts it into information that the management relies on it, in making management decisions in record time, which saves time, effort and cost. Likewise adds (Jeon et al., 2023; Fadhil et al., 2021) that artificial intelligence will push economic units to the stage of intellectual creativity through what it provides of data analysis and ideas for action. That applies to the blockchain or any other technique that grants the confidence to management for relying on digital data (Alnoor et al., 2022). This paper aims to rationalize management, by making decisions on investment derived from the information that accounting gives that results from meta-verse, by defining the appropriate accounting methods to deal with digital commercial transactions, and at the same time reaching the representation of events honestly, that will reflect the values of digital assets and give relevant information to users, this will reduces the gap between the virtual and real worlds (AL-Fatlawey et al., 2021). Hence the importance of this paper in highlighting on accounting recognition of digital assets and obligations, then carrying out the function of measurement and disclosure, and finally having the ability to produce reports at any time (Gatea and Marina, 2016). Thus, delivering information to users even when they are not present when people are unable to leave their homes as happened in the (Covid-19) pandemic.

2   Literature Review

2.1   Metaverse is a Digital Knowledge

The management might change the accounting methods applied in economic units, after the Fourth Industrial Revolution and for increasing reliance on digital technologies in various businesses through what accounting gives. After conducting a search for relevant studies, a study (Mystakidis, 2022; Manhal et al., 2023), emphasized what this technology provides in linking between virtual and real worlds to users. Furthermore, the virtual world is a digital environment that can be harnessed in the field of education. Additionally, in a study (Moll & Yigitbasioglu, 2019; Atiyah and Zaidan, 2022) the authors discuss how digital technologies are reflected in accounting work, represented by cloud data and big data with blockchain and artificial intelligence, and how to employ them in the accounting field as a result of their flexible nature, and what can be gained from automating economic decisions through digital technologies, then correct the financial vision towards the economic information received in a timely manner. Similarly, a study (Quinto, 2021, Atiyah, 2023), aimed to assess the ability of economic units to develop their business according to technology, and then obtain market benefits and achieve profits with optimal timings. The purpose of the research was to get new revenues, that help the unit overcome bankruptcy, while fulfilling promises towards launching high-quality products, via focusing on technology in their productivity in the future. Accountants will have a major role through the techniques helping them to process data and provide accounting information in a timely manner, then obtain user satisfaction and compete with similar units through information of high quality (Atiyah, 2020).

Another study dealt with augmented reality and its various uses in many professions, including industry, commerce represented by marketing, medicine, education, and others. So, through communication between the virtual and physical worlds, the user can be informed of existing services and the different kinds of industries, and explore digital product details without the need for touch (Muhsen et al., 2023; Al-Hchaimi et al., 2023; Chew et al., 2023). It is also possible to communicate with the economic unit to know the characteristics of the products the management's endeavors to develop them, and the possibility of measuring the management's performance in managing the unit's resources. This can be done through blockchain technology that processes a large amount of data in little time. Finally, financial reports can be made more flexible by providing users with various information, such as by adding pictures or video clips to financial reports (Mohammed, 2019).

A study conducted by (Ghani & Muhammad, 2019) about what employers expect from the skills of graduates, and the amount of technological knowledge they have about the market, then employ them in a way that serves economic units and develops their activity by using modern. Instead of the traditional methods of work, so it requires universities to educate the student in this direction to produce a student who knows the technology and masters it. Another research paper (Imene & Imhanzenobe, 2020) discusses the reflection of information technology on the work of accountants when preparing financial reports, undoubtedly, it is possible to reduce or eliminate errors by virtue of the help of technology. In addition to employing the internet service to store and display information instead of the traditional system that requires a long time, sometimes it might the client is always present in the workplace to inquire about his money not to mention, the economic unit's lack of an open labor market that reaches a larger segment. Which gives the unit more power in the labor market when using the technology. Also, the study (Al-Zoubi, 2017), explores the feasibility of the technology in reducing expenses and rapid access to information, and the openness that will happen to the unit within the local and global labor market. The study relied on a descriptive approach to collect relevant information from various studies on cloud computing, and the study concluded by identifying the important role of cloud computing in storing a huge amount of data to enable it to be recalled when needed. The study concluded that it is possible to control the size of the unit by reducing buildings and offices, to achieve the possibility of access to the financial system without the need for physical presence, as well as facilitating the conduct of economic operations and the preparation of financial reports.

Within the context of the same, the study showed that (Wang, 2023) the disclosure of the meta is not considered sufficient, except in the cases taking into account aspects of quality in the preparation and presentation of information. Furthermore, a study (Muravskyi et al., 2022) deals with accounting procedures related to digital economic transactions and the cryptocurrencies, to value digital assets that contribute to the contractual transaction. The research applied analytical methods were used to describe economic indicators for companies. The study concluded that the initial value is determined according to the economic transaction that exists, after that it is re-evaluated according to the fair value according to a future perspective.

2.2 Metaverse and Accounting

The main goal of an accounting job is to deliver information about the unit's resources, regardless of the method of transferring it to users whether by phone call, through the Internet, or the users visit personally the work site. The user might to be informed about the changes made to their money about the pervious period, sometimes the user being from other country, that's why economic units rely upon the digital aspect in delivering information to users. This will reflect on management's ability to take decisions at appropriate times, whether it is investment decisions or maintaining funding sources from various sources, like investors as well as obtaining loans from banks, this will reduce costs and spending little time studying the market. (Phornlaphatrachakorn & Kalasindhu, 2021). Accounting has a great role to play in helping everyone trade based on their financial orientation, by purchasing and trading in virtual assets that have no physical existence. This predicts the future to a promising future as a result of this investment trend in business, as we can see in Google site through the searches for the term meta-verse, to direct society in most of its categories, to find out the metaverse in many fields including accounting (Narin, 2021). Vyas (2021) confirmed blockchain technology that supports the metaverse allows individuals to create and invest in their digital assets, through the use of cryptocurrencies and non-fungible tokens such as images or videos which digital assets are linked to our real world, also, the possibility of promoting goods and services in an interactive manner which requires the importance of identifying relevant accounting methods to communicate information about the unit's resources to users, via finding accounting policies that are commensurate with the disclosure requirements in financial reports. The NFT market as a result of these technologies, reached sales of $2.5 billion in 2021 compared to $13.7 million in the previous year.

3 Hypothesis Development

3.1 Metaverse and Related Accounting Policies

A study conducted by (Pandey & Gilmour, 2023; Chayka, 2021) argue that the metaverse needs accounting practices different from what is commonly known, such as accounting for tax. Since deals take place in a virtual world other than reality, the question will be about what the appropriate accounting policy for financial reporting? The answer will be linked to the new accounting challenges imposed by the digital revolution, about how to do what might do. Giant companies working to keep pace with the system such as Apple and Meta and try to achieve the optimal investment benefit, for that economic units will need to organize their financial operations and record new business deals according to new accounting methods, consistent with the events through the meta-verse, for instill confidence between the parties to implement transactions. It is natural for users doubts at first about the ability of the economic unit into business (Jaber, 2022). Therefore, (Zadorozhnyi et al., 2023) provide evidence of the need for an accounting environment that deals objectively with this event, and keep the users trust of management actions through information they get. On the other hand, Pamungkas (2022) study dealt with the exploratory aspect of the metaverse by conducting qualitative research to examine the theoretical aspects of relevant studies, in order to reach an understanding of the future of business and investment that can be. Thus, the research hypothesis can be formulated as follows:

H1: Lack of accounting policies that help accountants find appropriate solutions to accounting issues arising from metaverse applications in economic units.

3.2 Metaverse and Its Relationship to Financial Reports

Today, the world is going through rapid digital changes, or what is called the fourth industrial revolution. These digital changes are racing with time which leads to a convergence of distance between the physical and digital world, as the picture between them converges in terms of the possibility of investing in these digital assets, via convergence through virtual reality and augmented reality (AR / VR) and the tools used for it. The method of accounting dealing with those assets that accountants might show in financial reports, poses a challenge to the accounting profession need to solve in light of the fourth industrial revolution, by providing a guide to accountants to do jobs without mistakes, with into account the relevant international guidelines and accounting directives. Therefore, an accounting culture might be generated that raises the accountant's awareness of what might be done that are related to digital assets, finally reflected in the quality of information then in the confidence of users and correct their decisions, by giving them reports that reflect the economic reality ultimately, this consistent with what was mentioned by (Carter, 2019; Al-Jazzar, 2023). Also, among the positive results that can be achieved when adopting the metaverse of economic units are information that will result from the data in the financial reports, reducing the time, increasing the accountant's culture towards digital development, and dealing with data on an automatic basis (Youssef & Attia, 2022). To reach these results, the characteristics of the metaverse and their reflection in the financial reports might be identified (Song et al., 2023). Thus, the research hypothesis can be formulated as follows:

H2: Lack of appropriate accounting methods for presenting digital assets in financial reports, when the economic unit exercises the digital business transaction.

4 Methodology

4.1 Sample Size and Measurements

The steps of accounting work in light of the fourth industrial revolution are no longer the same as their predecessors. Therefore, it is natural that in light of the continuous digital development, the accounting profession is required to take into account this, in order to meet the needs of users, in obtaining information commensurate with the huge amount of information and in line with the novelty this development. Therefore, accountants are required to deal in another way with economic resources of a digital nature. Via developing their accounting cultures in a way that is in line with the nature of the emerging economic event, by educating them and enabling them with digital transactions, then the possibility of disclosing those transactions in financial reports by the form of assets or liabilities, which suits different needs of users, which is finally reflected in the unit's ability to continue, also guarantee sources of financing and achieve gains. (Icaew, 2018). Therefore, the research sample will be the company (BVNK), which is a company that specialized in providing financial services, and the following (Table 2.) represents an overview of the history of its emergence and development.

The company has more than 200 members who finance an amount of $40 million, and it is supported by a number of partners (Fireblocks, Onfido, Comply Advantage, Chainalysis, Copper and B2C2). The company also electronic services as shown in the following (Fig. 3.):

Source: Preparation of the researcher based on: (BVNK, 2023; date of visit: 07–08-2023- 02:00 A.M.).

4.2 Accounting Business Model

Accordingly, the practical framework will depend on how to identify the digital asset and then measure and disclose it at last, with postponement of the relevant subsequent accounting events for the future. Also, postponement of revealing the nature of the revenues arising from those assets and the costs associated with these revenues. By take digital bill and meet the requirements with what Deloitte and IFRS offered in their reports, for what the required from the accountants do in such cases, based on the bills a company (BNVK) specializing in digital transactions. We will be established a working guide for the steps of accounting work in order to achieve the goals of the research, with the addition of clarifications by virtue of the analysis of the relevant conclusions, then make a financial position list that includes the research part in which the digital asset

Table 2. The history of the establishment of the BVNK company.

The date	The details
01–07-2021	Hold a license to operate as a FSCA approved Financial Services Provider in South Africa
01–08-2021	Starting working for the Cape Town office
	Continuation of the same schedule
01–11-2021	Starting working for the London office
01–04-2022	Starting with $40 million in Series A financing
01–10-2022	Obtaining the guarantee of registration of the Virtual Asset Service Provider by the Bank of Spain
01–11-2022	The company obtained additional funding from three offices amounting to 200 million dollars
01–11-2022	B2B (business-to-business) payments launched across the UK and Europe
01–12-2022	The company obtained an electronic money license from the United Kingdom

Source: (BVNK, 2023; date of visit: 03–08-2023- 12:10 A.M.)

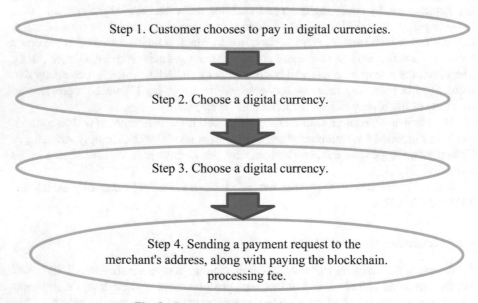

Fig. 3. Steps to conduct a digital transaction.

appears. Through the following (Fig. 4.), some types of digital assets can be presented, which will be the focus of accounting assets:

Source: (PWC, 2023: date of visit- On Friday, August 04, 2023 at 01:31 PM.).

Through (Trunfio & Rossi, 2022) and others, we obtained theoretical evidence for building an economic structure for units that activate with metaverse, with the aim of

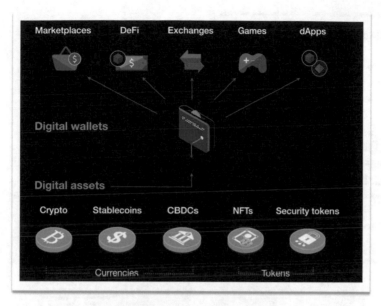

Fig. 4. Digital assets.

obtaining quality in providing economic information, this will support research paper towards accounting education through structure, to improve the culture of accountants and empower them in learning accounting methods in the shadow of the meta-verse, as a number of bodies interested in the world of metaverse and its applications issued narrative instructions, about what is required for accounting challenges that the accountants face. As a result of the challenges the metaverse has become the focus of attention of researchers through the increase in related studies In it, although the metaverse appeared in the 1950s of the last century (Talin, 2023). But the published studies increased from 2019 and 2021, which is the date of announcing the change of Facebook's name to Meta, not to mention the continuation of the studies until the date of writing this research paper.

In this paper, the quantitative approach will be used to identify the stages of accounting work, starting with the stage of accounting recognition, then the stage of accounting measurement to determine value to the asset according to the nature of accounting treatment, after that, procedure do the accounting disclosure and that the final step, that is the stage of communicating information about digital assets used when making economic decisions, that reflex on the interest of the unit and also maintains the funding required for work. When referring to previous relevant studies, what the researchers have done within the field of work has been reviewed to form an initial idea of the mechanisms of application, so it can a working structure will put in place that will be a new step toward dealing with digital resources, to deal with digital assets that lack a physical character which is agreed upon by studies the relevant. A working structure of several steps is presented which is a summary of the steps studied by the IFRS organization, Deloitte and EY as shown in the following (Fig. 5.):

Source: Preparation of the researcher based on (Deloitte, 202; IFRS, 2019).

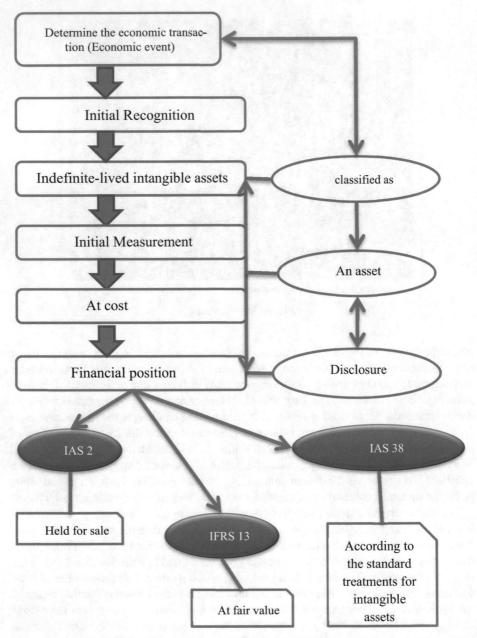

Fig. 5. Stages of accounting work for digital transactions

Through the above figure, a work form was put to carry out accounting work, starting from the stage of determining the economic transaction of digital, take into consideration of deferring the treatment of revenues and related costs, allowing future studies in this regard. Whether those were related to measuring the fair value or re-evaluation according to the accounting policies used by the unit that are compatible with the economic event. According to the model recognized by using invoices that include the digital asset, then followed by the accounting measurement process starting with the cost initially based on accounting recognition, and finally reaching the disclosure stage that can be made according to the asset kind, whether this asset is kept for normal activities or for the purpose of trading it represents a digital asset (Kapsis & Brown, 2015). As a result of these conditions and by applying the definition to the items of the statement of financial position, for reaching any of the items it is included the categorical It falls within the scope of digital assets, the following (Table 3) appears:

Table 3. The category of digital assets.

Classification	Recognition requirement	Reasons for recognition or not
Cash	It cannot be counted in cash or in exchange for cash	Lack of physical presence
Financial instrument or financial asset	It is not a contract that entails payment or receipt of cash	=
Inventory	It is not considered a tangible asset and therefore it is not considered an inventory	=
Intangible asset	Being intangible, it satisfies the definition of an intangible asset (other than a financial asset or goodwill)	It meets the classification within the intangible asset

Source: (KPMG, 2022; Cappiello et al., 2022)

When referring to the above table, it becomes clear that the digital assets did not meet any of the definitions of the items in the statement of financial position, as is the case with the items of cash or the item of financial assets and inventory. While digital assets met the definition of intangible assets after the aforementioned conditions, which were mentioned in the International Financial Reporting Standards were met. Therefore, the standards that deal with intangible assets can be adopted in terms of recognition and then accounting measurement, and it is possible to use other relevant standards if they are compatible with the transactions of intangible assets according to the standards adopted by the economic unit, as is the case in the IAS 8 standard, which allows the unit to use standards Which it deems appropriate in accordance with the economic event, conditional on compatibility with the nature of the intangible asset.

Among the assets that can be considered digital assets for example, a digital license for photos, video clips, tools used in games such as clothes, weapons, etc., also digital currencies in exchange for obtaining physical assets such as bonds or precious metals, can be controlled and obtained by the user via currencies which can be harnessed for such products (Muir, 2022). It is worth, noting that encoding the digital asset in light of the meta-verse, means does not repeat which prevents change or modification in the digital transaction, this leads to several benefits, including: (McKinney et al., 2023).

–   Faster transaction processing as in the blockchain that supports meta-verse.
–   Obtaining higher security by identifying the dealers in the transactions.
–   Reducing the costs associated with each transaction, by accelerating the processing
    of transactions and delivery of information to users in actuall time.

Also, (Jackson & Luu, 2023) classifies digital assets into two classifications, as
shown in the following (Fig. 6.):

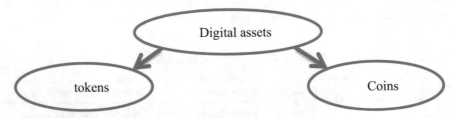

Fig. 6. Classifies digital assets.

Source: Prepare the researcher.

By using the search sample through the BVNK Company bill, a digital invoice can
be displayed as follows. After making the invoice and making a list of all operations in
a manner similar to the ledger records, except that it took place in an electronic way for
the digital operations group, by applying the form in the form (Fig. 7.), accountants will
be aware of what is required of appropriate accounting procedures for economic events
in light of the fourth industrial revolution, and what It is required of the accountant
when there is an economic transaction of a digital nature in the light of the meta-verse.
After taking the bills above from the company and through the balances for the set of
operations that took place, the following appears.

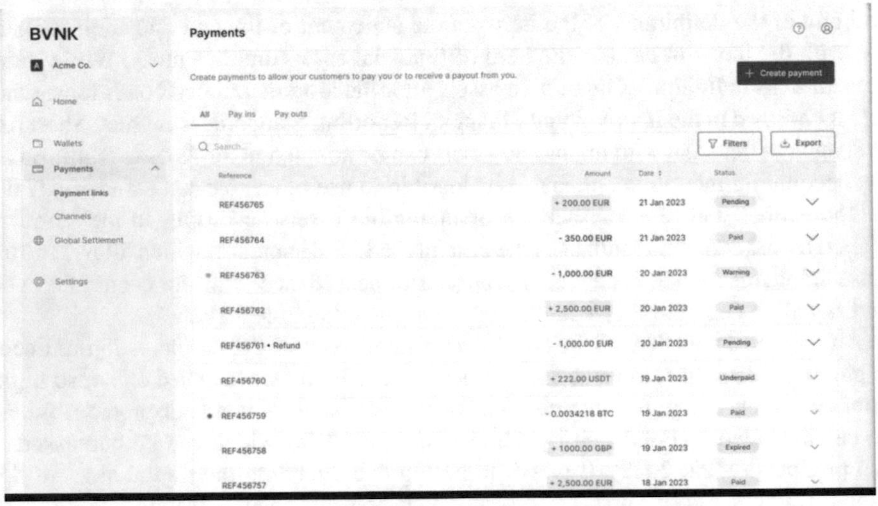

Fig. 7. Digital bills records.

Source: (BVNK, 2023; date of visit: 08–08-2023- 01:07 A.M.).

So, it will become logical for accountants to know what will happen to the digital asset at disclosure of an accounting procedure, to present in the financial statements, where it is measured according to the initial cost of the value of the digital transaction. Also, it will be presented in the statement of financial position under the item of intangible assets according to one of the criteria, that can be used based on the economic event.

5   Results

5.1   The Accounting Model for Digital Assets

Through the commercial transaction of the digital asset classified as an intangible asset is created, then the series of accounting steps shown in the (Fig. 8.) is applied, the features of the previously established accounting work become clearer.

The accounting recognition of the digital transaction as an intangible asset appears in the structure, then conducting accounting measurement at cost 200 EUR and disclosure in the statement of financial position via three options, because it meet the requirements for accounting recognition as an intangible asset, as it seems that the accountant has many options for dealing with such assets. The treatment methods for digital assets vary according to the activity of the unit, between considering the digital asset as an intangible asset or inventory according to the international standards used or else based on the transaction. It is also possible within the clarifications, disclosure might be made in the financial statements of digital assets, the purchases, and the sales, and the returns achieved also the results of economic transactions on the company's profits, what has been achieved at the level of the current year, and what the company expects from subsequent events, as well as future plans expected to be implemented. Under the item of intangible assets or inventory, according to the accounting policy applied, whether it is disclosure as diagrams, graphics, or texts that list to communicate information to users (Ferris, 2023), which is reflected in the decisions of users towards increasing the volume of investment or keeping it.

5.2   Accounting Disclosure of Digital Assets

Through the proposed model for the purpose of showing digital assets as intangible assets, it is possible to indicate the classification in which the digital assets will appear and under any item in the statement of financial position as shown in (Table 4.)Furthermore, the digital assets subject to the application of the standard adopted by the company in its commercial activity according to the company's policies relevant policies, commensurate with economic event, digital assets can be displayed through the following.

Shown in the image below kind of digital asset from the game (Decentraland-Land), will be bought or sold on the basis of digital currencies within metaverse technologies.

Although there are many questions related to the fair value and what is required to record. Additionally, there are accounting operations related to gains or losses, it was not possible to present these topics in full and they can be discussed in the future. (Fig. 9.) When we complete the accounting process from the step of determining the economic

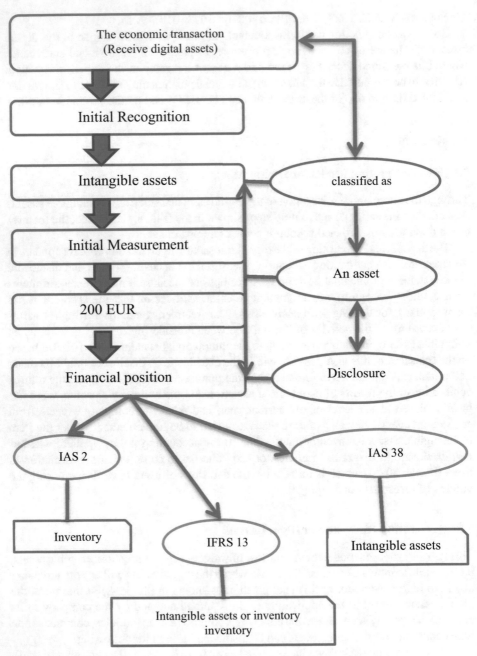

Fig. 8. The applied side of the accounting model.

event represented by the accounting recognition, then to the stage of preparing it for display in the financial statements, accordingly, through the following table it is possible

Table 4. Classification of the digital asset according to recognition.

Assets	Assets
Intangible Assets × × ×	Inventory × × ×

Continue Table 4

E.g.: (Crypto currencies/digital currencies, Utility tokens, Security tokens, Asset-backed tokens, Non-fungible tokens (NFTs) and Stable coins). (Jackson & Luu, 2023: 3–4)

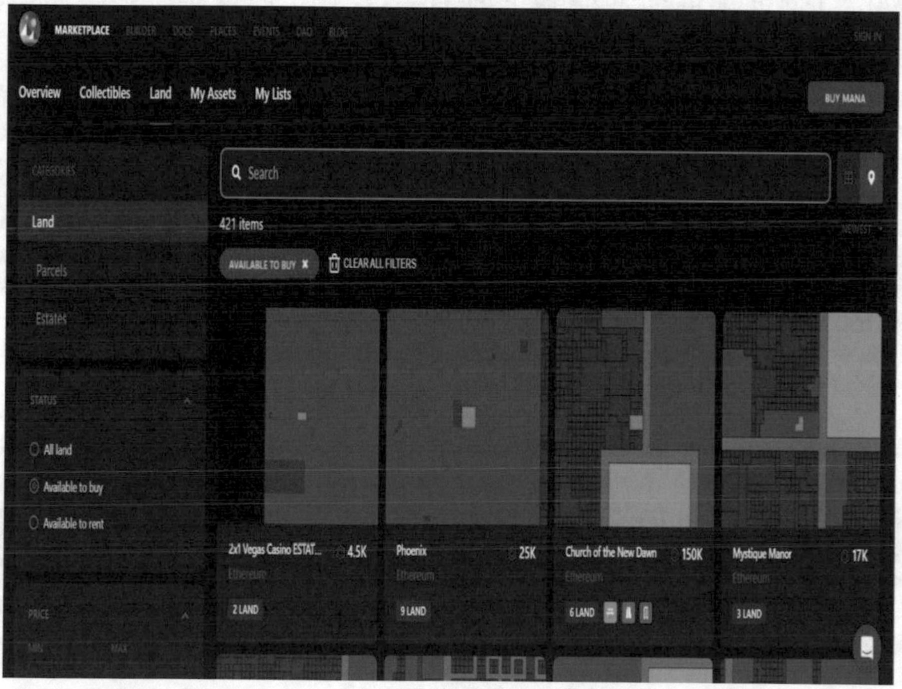

Fig. 9. Digital land. Source: (Decentraland, 2023; date of visit: 13–08-2023: 12:58 A.M.).

to display a trial balance that includes the part related to digital assets only, they are of the two types of inventory and intangible assets:

Table 5. Meet the conditions for recognition of the digital asset.

The definition	Meet the definition	Proof
1. Caused by a past event	Met within this condition	21–01-2023 The date of Purchase
2. It generates future benefits	Met within this condition	It can be bought and sold
3. It can be measured reliably	Met within this condition	It has a value of 200 EUR

Thus, it is evident that digital assets meet the definition according to the three conditions, which allows the result about the accounting recognition of those assets and their reporting in the financial reports. (Table 5.)

6 Discussion

The study was conducted to create rules for accounting work, that serves as a starting point for accountants in order to educate them about the requirements of accounting work in light of the fourth industrial revolution. Metaverse and its applications need to adopt accounting policies with the nature of digital transactions, because there are no clear mechanisms that help accountants perform their tasks efficiently. Furthermore, the importance of educating management via consistency in the application of accounting policies through accountants, in a way that leaves no room for erroneous judgment and its reflection on credibility and distortion in their reporting. Additionaly, the accountants face a huge amount of data and they need to convert it into real-time information to work in accordance on digital economic events, to learn how process the data and store it as a database for the full process. Through research in the literature, it is noted the extent of the wide interest in digital technologies, the blockchain is a ledger similar to the traditional ledger, which it is possible to get and maintain handlers of transactions without tampering in data, cloud computing is also a place for process and storage. (Rabih, 2023; Chukwunonso et al., 2022).

Thus, units can through the technology of the blockchain that supports metaverse and transitioning data between the virtual world and the real world in a way that helps to structure information and evaluate it for users. In order to overcome the challenges facing the accounting profession when conducting business transactions, the digital revolution requires accountants to be familiar with what is required of them. This research paper added new accounting practice according to a practical guide based on (IFRS, IASB, etc.). Our work guide the policymakers to develop and prepare the financial reporting that reflected on the information provided to users, for gaining their trust and finally preserving the funding sources, that necessary to operate the unit and ensuring its continuity, to keep pace with the ongoing digital change (Aharon et al., 2022).

7 Conclusion

Despite the huge development, the fears of the economic units cannot be resolved in the short term. Instead, building trust takes a long time, and this would create a responsibility for the accounting profession to overcome this barrier of fear. Therefore, a lot of responsibility falls on accountants towards the accounting tasks that these technologies contain, management need from their accountants to broaden their perceptions, then check the possibility of responding to these accounting techniques under any condition. It is not possible to access all accounting issues in this research, but it is possible to point out the importance of paying attention to artificial intelligence and its great role in changing the perception of business, especially since business activity is no longer limited to the borders of a country.

Through what has been explored in this paper for metaverse techniques and their various applications, it is clear that there is a major challenge facing accounting in the practice of accounting work and presentation of financial reports, with this huge amount of information that is communicated to users, professionals and those interested in accounting fields. As this guide is a step towards an accounting culture that helps accounting to process data and present it in the form of information to users. Thus avoiding any obstacles arising from commercial transactions in their digital format. As the research paper concluded, that there are a number of professional bodies with the competence to develop guidelines, recognizing that these accounting stages are accounting procedures with broad lines and information about digital assets can be presented in the form of pictures, narrative information, graphs, and other relevant methods. This would enable the accounting profession to give economic information suitable for the unit actions and provide the relevant information to users not to mention the truthfulness of that information's representation of commercial transactions, and setting the general rules for accounting work.

References

1. Abbas, S., et al. (2023). Antecedents of trustworthiness of social commerce platforms: a case of rural communities using multi Group SEM & MCDM Methods. Electronic commerce research and applications, 101322
2. Aharon, D.Y., Demir, E., Siev, S.: Real returns from unreal world? Market reaction to Metaverse disclosures. Res Int Bus Finance **63**, 101778 (2022). https://doi.org/10.1016/j.ribaf.2022.101778
3. Al-Abrrow, H., Fayez, A.S., Abdullah, H., Khaw, K.W., Alnoor, A., Rexhepi, G.: Effect of open-mindedness and humble behavior on innovation: mediator role of learning. Int. J. Emerg. Markets (2021)
4. Albahri, A.S., et al.: Based on the multi-assessment model: towards a new context of combining the artificial neural network and structural equation modelling. Rev. Chaos. Solitons Fractals **153**, 111445 (2021)
5. AL-Fatlawey, M. H., Brias, A. K., Atiyah, A. G. (2021). The role of Strategic Behavior in achievement the Organizational Excellence Analytical research of the manager's views of Ur State Company at Thi-Qar Governorate. J. Adm. Econ, 10(37).
6. Al Gnbri, M.K.D.: Internal auditing in metaverse world: between the prospects of virtual reality and the possibilities of augmented reality. Indonesian Acc. Rev. **12**(2), 125 (2022). https://doi.org/10.14414/tiar.v12i2.2848
7. Al-Gnbri, m.: Accounting and auditing in the metaverse world from a virtual reality perspective: a future research. J Metaverse **2**(1), 29–41 (2022)
8. Al-Hchaimi, A.A.J., Sulaiman, N.B., Mustafa, M.A.B., Mohtar, M.N.B., Hassan, S.L.B.M., Muhsen, Y.R.: A comprehensive evaluation approach for efficient countermeasure techniques against timing side-channel attack on MPSoC-based IoT using multi-criteria decision-making methods. Egypt. Inform. J **24**(2), 351–364 (2023)
9. Al-Jazzar, M. A. G., (2022). Augmented Reality as one of the effective solutions to strengthen the link between design and production. J Archit. Arts Humanit. - Volume VII - Issue Thirty-One

10. Alnoor, A., Tiberius, V., Atiyah, A. G., Khaw, K. W., Yin, T. S., Chew, X., & Abbas, S. (2022). How positive and negative electronic word of mouth (eWOM) affects customers' intention to use social commerce? A dual-stage multi group-SEM and ANN analysis. International Journal of Human–Computer Interaction, 1-30.https://doi.org/10.1080/10447318.2022.2125610

11. Alsalem, M.A., et al.: Rise of multiattribute decision-making in combating COVID-19: a systematic review of the state-of-the-art literature. Int. J. Intell. Syst. **37**(6), 3514–3624 (2022)

12. Al-zoubi, A. M., (2017). The Effect of Cloud Computing on Elements of Accounting Information System. Global J Manage. Bus. Res: DAccounting and Auditing, Vol 17 Issue 3 Version 1.0

13. Atiyah, A.G.: The effect of the dimensions of strategic change on organizational performance level. PalArch's J. of Archaeol. Egypt/Egyptology **17**(8), 1269–1282 (2020)

14. Atiyah, A. G. Strategic Network and Psychological Contract Breach: The Mediating Effect of Role Ambiguity.

15. Atiyah, A. G., zaidan, r. A. (2022). Barriers to using social commerce. In artificial neural networks and structural equation modeling: Marketing and Consumer Research Applications (pp. 115–130). Singapore: Springer Nature Singapore.

16. Baszucki, D., (2021). The Metaverse is coming. Available on: https://www.wired.co.uk/article/metaverse

17. Bozanic, D., Tešić, D., Puška, A., Štilić, A., Muhsen, Y.R.: Ranking challenges, risks and threats using fuzzy inference system. Decis. Making: Appl. Manag. Eng. **6**(2), 933–947 (2023)

18. BVNK, (2023). Our history. Available on: https://www.icaew.com/technical/technology/artificial-intelligence/artificial-intelligence-the-future-of-accountancy

19. BVNK, (2023). Send and receive digital assets. Available on: https://www.bvnk.com/payments

20. Cappiello et al., (2022). Accounting for and Disclosure of Crypto Assets. IASB Agenda reference 12A & FASB Agenda reference 12A

21. Carter, W., (2019). Defining the Technologies of the Fourth Industrial Revolution. Center for Strategic & International Studies, pp. (16–21). Available on: https://www.csis.org/analysis/beyond-technology-fourth-industrial-revolution-developing-world

22. Chayka, K., (2021). Facebook Wants us to live in the Metaverse, The New Yorker, Available on: https://www.newyorker.com/culture/infinite-scroll/facebook-wants-us-to-live-in-the-metaverse

23. Chew, X., Khaw, K. W., Alnoor, A., Ferasso, M., Al Halbusi, H., Muhsen, Y. R. (2023). Circular economy of medical waste: novel intelligent medical waste management framework based on extension linear Diophantine fuzzy FDOSM and neural network approach. Environ. Sci. Pollu. Res. 1–27

24. Chukwunonso et al., (2022). Security in Metaverse: A Closer Look. Available on: https://www.researchgate.net/publication/358948229_Security_in_Metaverse_A_Closer_Look

25. Dange, J., (2023). 5 Technologies That Are Powering the Metaverse. Available on: https://www.encora.com/insights/5-technologies-that-are-powering-the-metaverse

26. Decentraland, (2023). Available on: https://market.decentraland.org/lands

27. Deloitte, (2022). The Metaverse — Accounting Considerations Related to Non fungible Tokens. Available on: https://dart.deloitte.com/USDART/home/publications/deloitte/accounting-spotlight/2022/metaverse-accounting-considerations

28. Deloitte, (2023). Technology Industry Accounting Guide. Available on: https://dart.deloitte.com/USDART/home/publications/deloitte/industry/technology/technology-accounting-guide

29. Dwivedi, Y.K., et al.: Metaverse beyond the hype: Multidisciplinary perspectives on emerging challenges, opportunities, and agenda for research, practice and policy. Int. J. Inform. Manage. **66**, 102542 (2022). https://doi.org/10.1016/j.ijinfomgt.2022.102542
30. Eneizan, B., Mohammed, A.G., Alnoor, A., Alabboodi, A.S., Enaizan, O.: Customer acceptance of mobile marketing in Jordan: An extended UTAUT2 model with trust and risk factors. Int. J. Eng. Bus. Manage. **11**, 1847979019889484 (2019)
31. Fadhil, S.S., Ismail, R., Alnoor, A.: The influence of soft skills on employability: a case study on technology industry sector in Malaysia. Interdiscip. J. Inf. Knowl. Manag. **16**, 255 (2021)
32. Far, S.B., Rad, A.I.: Applying digital twins in metaverse: user interface, security and privacy challenges. J. Metaverse Rev. Art. **2**(1), 8–15 (2022)
33. Ferris, S., (2023). A taxonomy for classifying digital assets. Available on: https://www.journalofaccountancy.com/issues/2023/jul/a-taxonomy-for-classifying-digital-assets.html
34. Fortune Business Insights, (2023). Metaverse market size. Available on: https://www.fortunebusinessinsights.com/metaverse-market-106574
35. Gatea, A.A., Marina, V.: Higher education funding in Iraq in terms of the experience of particular developed countries. Int. J. Adv. Stud. **6**(1), 8–17 (2016)
36. Ghani, E. K. & Muhammad, K., (2019). Industry 4.0: Employers expectations of accounting graduates and its implications on teaching and learning practices. Int. J. Educ. Pract.: Vol. 7 No. 1 (2019)
37. Gibbs, A., (2021). What is the metaverse? The 101 Guide to the internet sucessor. PWC, Available on: https://www.pwc.com.au/digitalpulse/101-metaverse.html
38. Grand View Research, (2023). Metaverse Market Size, Share Trends Analysis Report By Product, By Platform, By Technology (Blockchain, Virtual Reality (VR) & Augmented Reality (AR), Mixed Reality (MR)), By Application, By End-use, By Region, And Segment Forecasts, 2023 - 2030. Report ID: GVR-4-68039-915-5. Available on: https://www.grandviewresearch.com/industry-analysis/metaverse-market-report
39. Grayscale, (2021). The Metaverse: Web 3.0 Virtual Cloud Economies. Available on: https://www.digitalcapitalmanagement.com.au/insights/the-metaverse-web-3-0-virtual-cloud-economies/
40. Hamid, R.A., et al.: How smart is e-tourism? a systematic review of smart tourism recommendation system applying data management. Comput. Sci. Rev. **39**, 100337 (2021)
41. Haouam, D.: IT governance impact on financial reporting quality using COBIT framework. Glob. J. Comput. Sci.: Theor. Res. **10**(1), 1–10 (2020). https://doi.org/10.18844/gjcs.v10i1.4143
42. ICAEW, (2018). Artificial intelligence and the future of accountancy. Center for Strategic & International Studies. Available on: https://www.icaew.com/technical/technology/artificial-intelligence/artificial-intelligence-the-future-of-accountancy
43. IFRS, (2019). Holdings of Cryptocurrencies. Available on: https://www.ifrs.org/content/dam/ifrs/supporting-implementation/agenda-decisions/2019/holdings-of-cryptocurrencies-june-2019.pdf
44. Imene, F. & Imhanzenobe, J., (2020). Information technology and the accountant today: What has really changed? J Account. Taxation, Vol. 12(1), pp. 48–60, January-March 2020
45. Jaber, T. A., (2022). Security Risks of the Metaverse World. Int. J. of Interact. Mob. Technol. (iJIM), VOL. 16 NO. 13
46. Jackson, A.B., Luu, S.: Accounting for digital assets. Aust. Account. Rev. (2023). https://doi.org/10.1111/auar.12402
47. Jeon, et al.: Blockchain and AI Meet in the metaverse. IntechOpen (2021). https://doi.org/10.5772/intechopen.99114
48. Kalyvaki, M.: Navigating the metaverse business and legal challenges: intellectual property, privacy, and jurisdiction. J. Metaverse **3**(1), 87–92 (2023). https://doi.org/10.57019/jmv.1238344

49. Kapsis, M. & Brown, J., (2015). Conceptual Framework Elements of financial statements—definitions and recognition. Available on: https://dart.deloitte.com/USDART/home/publicati ons/deloitte/accounting-spotlight/2022/metaverse-accounting-considerations

50. Khaw, K.W., et al.: Modelling and evaluating trust in mobile commerce: a hybrid three stage Fuzzy Delphi, structural equation modeling, and neural network approach. Int. J. Hum-Comput. Interact. **38**(16), 1529–1545 (2022)

51. KPMG, (2022). Accounting for crypto assets. Available on: https://frv.kpmg.us/reference-lib rary/2022/accounting-for-crypto-assets-by-investment-companies.html

52. Krishnamoorthy et al., (2020). Design and implementation of IoT based energy management system with data acquisition. IEEE 7th International Conference on Smart Structures and Systems ICSSS 2020. DOI: https://doi.org/10.1109/ICSSS49621.2020.9201997

53. Manhal, M., Al-khalidi, A., Hamad, Z.: Strategic network: managerial myopia point of view. Manage. Sci. Lett. **13**(3), 211–218 (2023)

54. McKinney et al., (2023). AICPA Updates Practice Aid on Digital Assets, and Other Crypto Accounting Hot Topics. Volume 30, Issue 6

55. Mohammed, T. I., (2019). Innovative methods in e-marketing with augmented reality and their impact on product design. J. Architect., Arts and Humanities, Issue seventeen

56. Moll, J., Yigitbasioglu, O.: The role of internet-related technologies in shaping the work of accountants: New directions for accounting research. British Acc. Rev. **51**(6), 100833 (2019). https://doi.org/10.1016/j.bar.2019.04.002

57. Muhsen, Y.R., Husin, N.A., Zolkepli, M.B., Manshor, N., Al-Hchaimi, A.A.J.: Evaluation of the routing algorithms for NoC-Based MPSoC: a fuzzy multi-criteria decision-making approach. IEEE Access **11**, 102806–102827 (2023). https://doi.org/10.1109/ACCESS.2023. 3310246

58. Muir, S., (2022). Accounting for NFTs. Available on: https://frv.kpmg.us/reference-library/ 2022/accounting-for-nonfungible-tokens-nfts.html

59. Mujiono, M.N.: The Shifting Role of Accountants in the Era of Digital Disruption. (2021). https://doi.org/10.11594/ijmaber.02.11.18

60. Muravskyi et al., (2022). Accounting and audit of electronic transactions in metaverses, Visnyk ekonomiky – Herald of Economics, 2, 128–141. DOI: https://doi.org/10.35774/visnyk2022. 02.128 .https://doi.org/10.35774/visnyk2022.02.128

61. Mystakidis, S.: Entry Metaverse. . Encyclopedia **2022**(2), 486–497 (2022). https://doi.org/ 10.3390/encyclopedia2010031

62. Narin, N.G.: A Content analysis of the metaverse articles. J. Metaverse Res. Articl **1**(1), 17–24 (2021)

63. Pamungkas, B., (2022). The future of cities in metaverse Era. Conference: The 4th Open Society Conference OSC 2022

64. Pandey, D., Gilmour, P.: Accounting meets metaverse: navigating the intersection between the real and virtual worlds. J. Financ. Reporting Account. (2023). https://doi.org/10.1108/ JFRA-03-2023-0157

65. PHORNLAPHATRACHAKORN, K., KALASINDHU, K. N., (2021). Digital accounting, financial reporting quality and digital transformation: evidence from thai listed firms, J Asian Finance, Econ. Bus., Vol 8 No 8 (2021) 0409–0419 409

66. PWC, (2022). Seeing is believing. Available on: https://www.pwc.com/seeingisbelieving

67. PwC, (2023). Demystifying cryptocurrency and digital assets. Available on: https://www. pwc.com/us/en/tech-effect/emerging-tech/understanding-cryptocurrency-digital-assets.html

68. Quinto, L. C. D., (2021). NFTs Augmented Reality - The development of a new business model. Open Repository of the Universities of Applied Sciences. Available on: https://www. theseus.fi/handle/10024/512910

69. Rabih, M. I., (2023). Accounting information system in the world of metaverse - an exploratory study. Sci. J Commercial Res. (Menoufia University), Article 11, Volume 50, Issue 3, July 2023, Page 537–578
70. Shein, E., (2022). 7 top technologies for metaverse development. Available on: https://www.techtarget.com/searchcio/tip/7-top-technologies-for-metaverse-development
71. Solulab, (2022). Metaverse Platforms. Available on: https://www.solulab.com/metaverse-development-company
72. Song, et al.: Exploring the key characteristics and theoretical framework for research on the metaverse. Appl. Sci. **2023**(13), 7628 (2022). https://doi.org/10.3390/app13137628
73. Talin, T., (2023). History and Evolution of the Metaverse Concept. Available on: https://morethandigital.info/en/history-evolution-of-metaverse-concept/
74. Trunfio, M., Rossi, S.: Advances in metaverse investigation: streams of research and future agenda. Virtual Worlds **1**(2), 103–129 (2022). https://doi.org/10.3390/virtualworlds1020007
75. Vyas, N. G., (2021). What is the Metaverse and How Do Enterprises Stand to Benefit? Available on: https://www.itbusinessedge.com/networking/metaversc-enterprises-benefits/#How_Does_the_Metaverse_Work
76. Wang, Y.: Voluntary Information Disclosure in Meta Verse Industry——Taking IMS Group Information Disclosure Violation as Case Analysis. In: Mallick, H., Gaikar, V.B., San, O.T. (eds.) Proceedings of the 2022 4th International Conference on Economic Management and Cultural Industry (ICEMCI 2022), pp. 1728–1739. Atlantis Press International BV, Dordrecht (2023). https://doi.org/10.2991/978-94-6463-098-5_195
77. Youssef, M. M. A., Attia, N. S. M., (2022). A proposed introduction to using Metaverse technology as one of the information technology innovations in improving the quality of financial reports in the Egyptian environment "between the determinants of use ... and the advantages and risks of application", Conference: Challenges and prospects for the accounting and auditing profession in the twenty-first century. Available on: https://www.researchgate.net/publication/361250939
78. Zadorozhnyi, Z.-M., Muravskyi, V., Humenna-Derij, M., Zarudna, N.: Innovative accounting and audit of the metaverse resources. Mark. Manage. Innovations **13**(4), 10–19 (2022). https://doi.org/10.21272/mmi.2022.4-02

The Effect of Digital Business Strategy on Improving Customer Journey: Evidence from Users of Virtual Reality Platforms

Mohammad Abd Al-Hassan Ajmi Al-Eabodi[✉] and Ammar Abdulameer Ali Zwain

Faculty of Administration and Economics, University of Kufa, Kufa, Iraq
mzmitab@gmail.com, ammara.zwain@uokufa.edu.iq

Abstract. The current research aims to test the effect of digital business strategy in improving the customer journey. This study targeted users of virtual reality platforms, namely Roblox and Facebook. The sample size was 126 users. In addition, structural equation modeling (PLS-SEM) method was used. The results of the research showed acceptance of the main hypothesis that there is an impact of the digital business strategy in improving the customer journey.

Keywords: Digital Business Strategy · Customer Journey · Virtual Reality

1 Introduction

In light of the rapid developments in the business world, many concepts have emerged that could affect the work of organizations in various sectors, including what is internal and what is external (Ahmed et al., 2023; Hadi et al., 2019). The financial sector and its financial institutions, including users of virtual reality platforms, are not isolated from being affected by these developments. It should be noted here that one of these important concepts at the present time is what is known as the "customer journey" (Purmonen et al., 2023). The customer journey is a fundamental concept in marketing and customer relationship management (Lundin & Kindström, 2023). It is seen as representing a series of steps and interactions that the customer undertakes from the moment, discovers the organization's product or service until obtains it and provides feedback after using it. Comprehensively and accurately understanding this journey represents a strategic challenge for business organizations to improve customer experience and increase customer satisfaction and loyalty. In order to achieve the main purpose of the research, a set of objectives were formulated, which are as follows: a) Determine the level of adoption of the digital business strategy by users of virtual reality platforms. b) Exploring the extent to which users of virtual reality platforms seek to improve the customer journey. c) Testing the effect of digital business strategy on improving customer journey in users of virtual reality platforms (Zhang et al., 2022).

Moreover, since the virtual reality platforms industry has witnessed a major shift towards the digital and technological orientation, as online services and platforms applications have become an essential part of the customer's journey (Krasonikolakis & Chen,

M. Al-Emran et al. (Eds.): IMDC-IST 2024, LNNS 895, pp. 212–222, 2023.
https://doi.org/10.1007/978-3-031-51716-7_14

2023). Hence, the digital business strategy is a vital tool in enhancing the customer experience and enabling them to access virtual reality platforms easily and achieve a high level of transparency and interaction (Hoang & Tan, 2023). It is worth noting that customers appreciate the smooth, personalized and customized experience, as they expect a distinguished and innovative experience from virtual reality platforms (Purmonen et al., 2023). Therefore, virtual reality platforms must invest in improving the customer's journey by applying the latest digital technologies, artificial intelligence and big data analysis, to achieve a unique and satisfactory user experience (Muhsen et al., 2023; Al-Hchaimi et al., 2023; Chew et al., 2023a). Previous studies neglected to study customer satisfaction and loyalty to the organization and its services (Wang et al., 2020; Priyanto et al., 2023). With regard to the variable (customer journey), the majority of previous studies were theoretical papers such as (Shen et al., 2020; Rita et al., 2023; Manhal et al., 2023). Hence, there was a lack of capture causal relationship. Hence, the current research contributes to bridging the knowledge gap. Hence, the current study provides a new insight and a comprehensive analysis of the mutual effects between these variables.

2   Literature Review

2.1   The Concept of Digital Business Strategy

The concept of digital business strategy is a recent topic. Mithas and Lucas who first coined the term digital business strategy published the first publication referring to digital business strategy as the fusion of IT strategy and business strategy in 2010. It was later expanded by Bharadwaj et al. (2013), who are starting to mark the topic and explain that digital business strategy is beginning to gain traction among researchers (Kahre et al., 2017; Fredericks, 2020; Uhlig & Remané, 2022; Alnoor et al., 2022). Scholars see the field of digital business strategy as a combination of management information systems and strategic management research (Bharadwaj et al., 2013). According to a group of researchers such as the digital business strategy represents a combination of organizational strategy and information technology strategy. This means that IT strategy becomes more important and must be seen as more than just a functional level strategy (Albayati et al., 2023).

In the same context, digital business strategy is evident as one of the most prominent concepts in the business world and is referred to as the intersection of strategic management and information technology (Holotiuk & Beimborn, 2017; Atiyah et al., 2022). It includes digital technology that is commonly integrated into systems in the current time (cloud computing, artificial intelligence, business intelligence, big data and similar technology) that help organizations excel and create value (Iafrate, 2018; Atiyah, 2022; Gatea and Marina, 2023). In contrast, Chi et al. (2016) argued that the digital business strategy is a strategy at the organizational level and not an information technology strategy or a functional strategy, because the purpose of including this strategy is to generate value for the business through the means of technology. This view was supported by the study of Kahre et al. (2017), where it was highlighted that the digital business strategy restructures the business model of organizations leading to digital transformation and is more concerned with producing potential financial results for organizations (Lee et al., 2023; Khaw et al., 2023). In addition, a study of Nadeem et al. (2018), explained the digital business strategy leads to digital transformation, yet the capabilities of the organizations that contribute to this are ambiguous. In this regard, Bataineh et al. (2015) argued that the ability of information technology for organizations is to enhance business efficiency regardless of the industry in which they operate (Abbas et al., 2023; Bozanic et al., 2023).

Bharadwaj et al. (2013) defines a digital business strategy as an organizational strategy that is formulated and implemented by utilizing digital resources to create differential value. While (Lisienkova et al., 2022) defined it as a means to achieve the goals of the organization through the introduction of digital tools or elements such as the Internet of Things, business intelligence systems, the use of big data, and others. Ukko et al. (2019) confirmed the digital business strategy consists of two dimensions.

a. Managerial Capabilities: Managerial capabilities refer to the abilities of managers to use digital tools and methods in business strategy, mindsets and skills of employees, as well as in the workplace to achieve organizational goals (Heubeck, 2023).
b. Operational Capabilities: Operational capabilities refer to an organization's ability to integrate digital methods, tools, and programs into the overall business processes and strategies of the organization (Zhu & Jin, 2023).

2.2 The Concept of Customer Journey

Since the competitive advantage of the organization is affected by customer satisfaction, a kind of thinking has begun to maintain the competitive advantage of the organization, through the philosophy of customer-oriented management. However, until the 1990s, service processes were not sufficiently studied, because most service research focused on the service provider rather than the customer (Schneider & Bowen, 1993; Sadaa et al., 2023). Accordingly, managers and researchers changed the classic service plan model, turning it into a customer-oriented tool that visually describes the service concept and operations from the customer's point of view. In 1999, Tseng and her colleagues introduced for the first time a new framework known as the "customer journey". By creating an innovative tool to improve service operations, through objective planning of customer service experience (Wylde et al., 2023; Tan et al., 2023; Chew et al., 2023b). This was the first introduction of the term "customer's journey" in the literature (Tseng et al., 1999). The customer journey has been extensively investigated in studies on service management and design (Mashhady et al., 2021; Atiyah, 2020b). While the service design literature consists of several concepts and tools, the customer journey is one of many widely implemented tools that have been used to understand complex customer actions and gain insights into their experience with the organization. However, it should be noted that customer satisfaction stems not only from the product or service, but also from the entire process or customer journey (Campiranon, 2022).

Van Vaerenbergh et al. (2019) explained that one of the main components of the customer journey is customer interactions with service providers. These are commonly known as customer journey touchpoints. In order for organizations to increase interaction with customers, they must monitor different touchpoints during the customer journey (Boyd et al., 2019). Touch points were first described in the scientific literature as encounters between service providers and customers. Lockwood and Jones (1989) described these encounters as interactive variables, specifically "personal characteristics, perceptions of each other, social competence, needs and goals" between clients and service providers. In the 1990s, researchers shed light on the social perception of such encounters with regard to service providers, liaison personnel, and customers (Czepiel, 1990). They focused on the quality factors affecting the said encounters during the service experience stages (Danaher & Mattsson, 1994; Atiyah, 2022; Sak et al., 2023).

There were many opinions of writers and researchers regarding the concept of the customer journey, and there was some difference in their views on this subject, as it was viewed from an interactive point of view, and from a phased point of view and behavioral at other times. Norton & Pine (2013) defined the customer journey as the sequence of events - whether designed or not - that customers go through to learn about, purchase and interact with an organization's offerings - including goods, services or experiences. Rana et al. (2023) defined it as a complete interaction between the customer and the organization from obtaining product and service information to making the final purchase decision, repurchase and more.

The customer journey is a relatively new term. The first references to this concept were made by Tseng et al. (1999), which they introduced to improve service operations, by mapping and modifying the service experience of customers from the customer's perspective. According to Campiranon (2022), the customer journey consists of the following dimensions:

a. Service satisfaction: Is a measure that determines the extent to which customers are satisfied with an organization's services and capabilities (Andalas, 2022).
b. Service failure and recovery: The process used to "restore" dissatisfied customers by identifying the problem and fixing it or compensating for a service failure (Anwar & Ozuem, 2022).
c. Co-creation: Refers to the process of designing a product or service in which input from customers plays a central role from start to finish. It also refers to any way in which an organization allows customers to submit ideas, designs, or content for a particular product or service (Pham et al., 2022).

Customer response: The set of positive or negative feedback an organization receives about its products, services, or business ethics (Nawaz et al., 2022). Based on the mentioned discussion, the hypothetical research model was developed, illustrating the nature of the relationship and the impact between the research variables, as shown in Fig. 1. In order to achieve the aforementioned research objectives and based on the hypothetical research model, the following research hypothesis was assumed:

H1: There is a significant effect of digital business strategy on improving customer journey.

3 Methodology

To discover the relationship between variables, we utilized survey approach. Questionnaire technique is a tool to data gathering. The users of virtual platforms such as (Facebook, Roblox) are sample data. As sample size was (126) user. Based on previous literature, a questionnaire was developed to measure the study variables. The scale of

Cui et al. was adopted. (2022) to measure customer journey while AlNuaimi et al.'s scale was used. (2022) to measure Digital Business Strategy. The common issue of bias is a growing problem in human research. In addition, Harman's single-factor analysis was used, and the results indicated that the variance rate was less than 50%. Regarding the characteristics of the sample, the percentage of males was 44%, while the percentage of females was 56%. Regarding age, the largest percentage was for the 30 to 40 years group, amounting to 39%. With regard to experience, the percentage of more than 10 years was the largest and amounted to 41%.

4 Data Analysis

The results of the descriptive analysis of the dimensions of digital business strategy show that there are various levels of prevalence of these dimensions in virtual platforms, based on the data provided. The relative importance of the dimensions shows different and close levels and values. Based on the results shown in Table 1, it can be concluded that the operational capabilities dimension is the most widespread, as the arithmetic mean for it was (3.95), with a standard deviation of (.760). As for the managerial capabilities dimension, it was the least prevalent with an arithmetic mean of (3.83), and a standard deviation of (.730).

Table 1. Descriptive analysis of digital business strategy

No.	Dimensions of Digital Business Strategy	Mean	Std.	R. Imp.	Sequence
1	Managerial Capabilities	3.83	.760	77%	2
2	Operational Capabilities	3.95	.730	79%	1
Total of Digital Business Strategy		3.89	.745	78%	Second

It is also worth noting that the digital business strategy variable ranked second among the study variables, with an arithmetic mean of (3.89), a standard deviation of (.745), and a relative importance of (78%), according to the results of Table 1.

The results of the descriptive analysis of the dimensions of customer journey show that there are various levels of prevalence of these dimensions in virtual reality platforms, based on the data provided. The relative importance of the dimensions shows different and close levels and values. Based on the results shown in Table 2, it can be concluded that the service failure and recovery dimension is the most prevalent, as the arithmetic mean for it was (3.98), with a standard deviation of (.812). As for the dimension of satisfaction with the service, it was the least prevalent with a mean of (3.82), and a standard deviation of (.797).

It is also worth noting that the entrepreneurial culture variable ranked first among the four variables of the study, with an arithmetic mean of (3.91), and a standard deviation of (.808), with a relative importance of (78%), according to the results of Table 2.

In order to test this hypothesis, we used SmartPLS3 program, and built a structural model in order to show the relationship path of the study variables represented by (digital

Table 2. Descriptive analysis of customer journey

No.	Dimensions of Customer Journey	Mean	Std.	R. Imp.	Sequence
1	Service Satisfaction	3.82	.797	76%	4
2	Service Failure and Recovery	3.98	.812	80%	1
3	Co-creation	3.94	.751	79%	2
4	Customer Response	3.89	.871	78%	3
Total of Customer Journey		3.91	.808	78%	First

business strategy) as an independent variable, and (customer journey) as a dependent variable, as shown in Fig. 1.

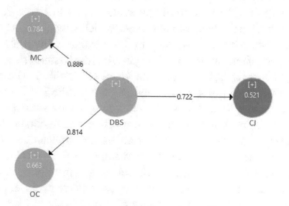

Fig. 1. Testing the research hypothesis

By looking at Fig. 2 of the structural equation model related to testing the main hypothesis of the research, and Table 3 of the results of the path analysis, the results show that the main hypothesis of the research can be accepted, which states that: "There is a significant effect of digital business strategy (DBS) on improving customer journey (CJ)". Where the value of the effect coefficient (β) reached (.722), and this effect is significant because the level of significance was (.000), which is less than the level of significance assumed by the researcher (0.05). In addition, the digital business strategy variable explains (52.1%) of all changes that can occur in customer journey variable, a value indicated by the Explanation Coefficient (R2). To make sure that the hypothesis was accepted or not, the (T) value test was implemented, and after analyzing it, the (T) value reached (16.606), which is greater than its tabular value of (2.00). Based on these results, the main hypothesis of the research is accepted.

Table 3. Results of the analysis of the main research hypothesis

Path	β	R^2	T. Value	Sig	Result
Digital Business Strategy → Customer Journey	.722	.521	16.606	.000	Accept

5   Data Analysis

The results found that there is a relatively high extent of adoption of the digital business strategy variable and its dimensions in the virtual reality platforms. The arithmetic means for this variable, which was extracted by relying on the (SPSS V.22) program, was (3.89), which indicates a relatively high degree of progress for virtual reality platforms in their adaptation to technological developments and digital transformations in the business environment. This progress helps it to improve the efficiency of operations and achieve cost savings through the transition to digital systems and technologies. Thus, the first objective of the research objectives was achieved, which stipulated, "Determine the level of adoption of the digital business strategy by the virtual reality platforms." There is a relatively high degree of endeavor towards improving the customer's journey on the part of virtual reality platforms, but it is not at the level of ambition. The arithmetic means of this variable, which was extracted by relying on the (SPSS V.22) program, reached (3.91), which indicates a great focus on developing customer experience by virtual reality platforms. This result reinforces the importance of understanding the stages of the customer journey and better meeting their needs and expectations. By improving the customer journey, virtual reality platforms can improve their level of satisfaction and loyalty, which contributes to increasing market share and promoting sustainable growth. Thus, the second objective of the research objectives has been achieved, which stipulated, "Exploring the extent to which virtual reality platforms seek to improve the customer journey." It was found that there is a significant effect of the digital business strategy in improving the customer journey in virtual reality platforms. This is done by relying on the value of the influence coefficient (β), extracted by relying on the (SmartPLS3) program, which amounted to (.722), which is positive and significant, since the achieved level of significance amounted to (.000), which is less than the level of significance assumed by the researcher. And amounting to (0.05). Notably, this result is consistent with the findings of (Wielgos et al., 2021) in their study. Based on these results, the third objective of the research was achieved, which stipulated, "Testing the effect of digital business strategy on improving customer journey in the virtual reality platforms."

6   Conclusion

In conclusion, this research sheds light on the effect that a digital business strategy can play in improving the customer journey in today's dynamic business landscape. Therefore, the research aimed to test the effect of digital business strategy in improving customer journey in virtual reality platforms. The results of this research show that virtual reality platforms that adopt an effective digital business strategy experience significant improvements in the customer journey. Digital tools and technologies can revolutionize the way virtual reality platforms interact with their customers, offering customized, seamless and convenient experiences across various touch points. As a result, customers feel more engaged, satisfied and loyal to the brand. Virtual reality platforms that strategically implement digital solutions, such as mobile apps and chatbots powered by artificial intelligence and data analytics are seeing higher customer satisfaction rates. These digital elements enable virtual reality platforms to anticipate customer needs, solve problems instantly and provide customized solutions. Finally, this research demonstrated that a digital business strategy is vital for virtual reality platforms seeking to elevate their customer journey, and leveraging digital technology will remain a critical factor in creating meaningful and lasting customer experiences.

References

Abbas, S., et al.: Antecedents of trustworthiness of social commerce platforms: a case of rural communities using multi group SEM & MCDM methods. Electron. Commer. Res. Appl. **62**, 101322 (2023)

Ahmed, M.G., Sadaa, A.M., Alshamry, H.M., Alharbi, M.A., Alnoor, A., Kareem, A.A.: Crisis, resilience and recovery in tourism and hospitality: a synopsis. In: Tourism and hospitality in Asia: crisis, resilience and recovery, pp. 3–19. Singapore: Springer Nature Singapore (2023). https://doi.org/10.1007/978-981-19-5763-5_1

Al Mazrooei, A.K., Hatem Almaki, S., Gunda, M., Alnoor, A., Manji Sulaiman, S.: A systematic review of K–12 education responses to emergency remote teaching during the COVID-19 pandemic. Int. Rev. Educ. **68**(6), 811–841 (2022)

Albayati, H., Alistarbadi, N., Rho, J.J.: Assessing engagement decisions in NFT Metaverse based on the Theory of planned behavior (TPB). Telematics Inform. Rep. **10**, 100045 (2023)

AL-Fatlawey, M.H., Brias, A.K., Atiyah, A.G.: The role of Strategic Behavior in achievement the Organizational Excellence "Analytical research of the manager's views of Ur State Company at Thi-Qar Governorate". J. Adm. Econ. **10**(37) (2021)

Al-Hchaimi, A.A.J., Sulaiman, N.B., Mustafa, M.A.B., Mohtar, M.N.B., Hassan, S.L.B.M., Muhsen, Y.R.: A comprehensive evaluation approach for efficient countermeasure techniques against timing side-channel attack on MPSoC-based IoT using multi-criteria decision-making methods. Egyptian Inform. J. **24**(2), 351–364 (2023)

Aljabhan, B.: Economic strategic plans with supply chain risk management (SCRM) for organizational growth and development. Alex. Eng. J. **79**, 411–426 (2023)

Alnoor, A., et al.: How positive and negative electronic word of mouth (eWOM) affects customers' intention to use social commerce? A dual-stage multi group-SEM and ANN analysis. Int. J. Hum.-Comput. Interact. 1–30 (2022)

AlNuaimi, B.K., Singh, S.K., Ren, S., Budhwar, P., Vorobyev, D.: Mastering digital transformation: the nexus between leadership, agility, and digital strategy. J. Bus. Res. **145**, 636–648 (2022)

Andalas, M.R.: Influence of internal service quality, employee satisfaction, external service satisfaction and customer satisfaction toward customer loyalty of Gojek service users in Malang. IPTEK J. Proc. Scr. **1**, 381–387 (2022)

Anwar, S., Ozuem, W.: An integrated service recovery process for service failures: insights from systematic review. J. Cetacean Res. Manag. **25**(4), 433–452 (2022)

Atiyah, A.G.: Impact of knowledge workers characteristics in promoting organizational creativity: an applied study in a sample of Smart organizations. PalArch's J. Archaeol. Egypt/Egyptology **17**(6), 16626–16637 (2020)

Atiyah, A.G.: The effect of the dimensions of strategic change on organizational performance level. PalArch's J. Archaeol. Egypt/Egyptology **17**(8), 1269–1282 (2020)

Atiyah, A.G.: Effect of temporal and spatial myopia on managerial performance. J. La Bisecoman **3**(4), 140–150 (2022)

Atiyah, A.G., Zaidan, R.A.: Barriers to using social commerce. In: Artificial Neural Networks and Structural Equation Modeling: Marketing and Consumer Research Applications, pp. 115–130. Singapore: Springer Nature Singapore (2022). https://doi.org/10.1007/978-981-19-6509-8_7

Baqer, N.S., et al.: Indoor air quality pollutants predicting approach using unified labelling process-based multi-criteria decision making and machine learning techniques. Telecommun. Syst. **81**(4), 591–613 (2022)

Bataineh, A., Alhadid, A., Alabdallah, G., Alfalah, T., Falah, J., Idris, M.: The role of information technology capabilities in capitalizing market agility in Jordanian telecommunications sector. Int. J. Acad. Res. Bus. Soc. Sci. **5**(8), 90–101 (2015)

Bharadwaj, A., Sawy, O.A., Pavlou, P.A., Venkatraman, N.: Digital business strategy: toward a next generation of insights. MIS Q. **37**(2), 471–482 (2013). https://doi.org/10.25300/MISQ/2013/37:2.3

Boyd, D.E., Kannan, P.K., Slotegraaf, R.J.: Branded apps and their impact on firm value: a design perspective. J. Mark. Res. **56**(1), 76–88 (2019)

Bozanic, D., Tešić, D., Puška, A., Štilić, A., Muhsen, Y.R.: Ranking challenges, risks and threats using fuzzy inference system. Decis. Making: Appl. Manage. Eng. **6**(2), 933–947 (2023)

Campiranon, K.: Enhancing the customer journey during COVID-19 through service design: a case study of pawnshops in Bangkok. Thammasat Rev. **25**(1), 124–144 (2022)

Chew, X., Alharbi, R., Khaw, K.W., Alnoor, A.: How information technology influences organizational communication: the mediating role of organizational structure. PSU Res. Rev. (2023)

Chew, X., Khaw, K.W., Alnoor, A., Ferasso, M., Al Halbusi, H., Muhsen, Y.R.: Correction to: circular economy of medical waste: novel intelligent medical waste management framework based on extension linear diophantine fuzzy FDOSM and neural network approach. Environ. Sci. Pollut. Res. Int. **30**(24), 66428 (2023)

Chew, X., Khaw, K.W., Alnoor, A., Ferasso, M., Al Halbusi, H., Muhsen, Y.R.: Circular economy of medical waste: novel intelligent medical waste management framework based on extension linear Diophantine fuzzy FDOSM and neural network approach. Environ. Sci. Pollut. Res. 1–27 (2023)

Chi, M., Zhao, J., Li, Y.: Digital business strategy and firm performance: the mediation effects of E-collaboration capability. In: The Fifteenth Wuhan International Conference on EBusiness: WHICEB 2016 Proceedings (Vol. 10, pp. 123–139). Inderscience Enterprises Ltd. (2016)

Cui, X., Xie, Q., Zhu, J., Shareef, M.A., Goraya, M.A.S., Akram, M.S.: Understanding the omni channel customer journey: the effect of online and offline channel interactivity on consumer value co-creation behavior. J. Retail. Consum. Serv. **65**, 102869 (2022)

Czepiel, J.A.: Service encounters and service relationships: implications for research. J. Bus. Res. **20**(1), 13–21 (1990)

Danaher, P.J., Mattsson, J.: Cumulative encounter satisfaction in the hotel conference process. Int. J. Serv. Ind. Manag. **5**(4), 69–80 (1994)

Fredericks, J.: Towards an understanding of the boundaries and characteristics of a digital business strategy. (Doctoral dissertation, Department of Information Systems, University of Cape Town) (2020)

Gatea, A.A., Marina, V.: Higher education funding in Iraq in terms of the experience of particular developed countries. Int. J. Adv. Stud. **6**(1), 8–17 (2016)

Hadi, A.A., Alnoor, A., Abdullah, H., Eneizan, B.: How does socio-technical approach influence sustainability? Considering the roles of decision making environment. Application of Decision Science in Business and Management 55 (2019)

Heubeck, T.: Managerial capabilities as facilitators of digital transformation? Dynamic managerial capabilities as antecedents to digital business model transformation and firm performance. Digital Bus. **3**(1), 100053 (2023)

Hoang, Ha., TanLe, T.: Heliyon **9**(9), e19719 (2023). https://doi.org/10.1016/j.heliyon.2023. e19719

Holotiuk, F., Beimborn, D.: Critical success factors of digital business strategy. In: 13th International Conference on Wirtschaftsinformatik, Frankfurt School of Finance & Management, Frankfurt, Germany (2017)

Iafrate, F.: Artificial Intelligence and Big Data: The Birth of a New Intelligence. Wiley (2018). https://doi.org/10.1002/9781119426653

Kahre, C., Hoffmann, D., Ahlemann, F.: Beyond business-IT alignment-digital business strategies as a paradigmatic shift: a review and research agenda. In: Proceedings of the 50th Hawaii International Conference on System Sciences, pp. 4706–4715 (2017)

Khaw, K.W., Camilleri, M., Tiberius, V., Alnoor, A., Zaidan, A.S.: Benchmarking electric power companies' sustainability and circular economy behaviors: using a hybrid PLS-SEM and MCDM approach. Environ. Dev. Sustain. **25**(2), 1–39 (2023)

Krasonikolakis, I., Chen, C.H.S.: Unlocking the shopping myth: can smartphone dependency relieve shopping anxiety?–A mixed-methods approach in UK omnichannel retail. Inf. Manage. **60**(5), 103818 (2023)

Lee, C.T., Ho, T.Y., Xie, H.H.: Building brand engagement in metaverse commerce: the role of branded non-fungible Toekns (BNFTs). Electron. Commer. Res. Appl. **58**, 101248 (2023)

Lisienkova, T., Chelekova, E., Mitrofanova, I., Shindina, T., Titova, J.: An approach for increase of digital/IT innovation adoption for sustainable development of transportation system. Transp. Res. Procedia **63**, 564–574 (2022)

Lockwood, A., Jones, P.: Creating positive service encounters. Cornell Hotel Restaurant Adm. Q. **29**(4), 44–50 (1989)

Lundin, L., Kindström, D.: Digitalizing customer journeys in B2B markets. J. Bus. Res. **157**, 113639 (2023)

Manhal, M., Al-khalidi, A., Hamad, Z.: Strategic network: managerial myopia point of view. Manage. Sci. Lett. **13**(3), 211–218 (2023)

Mashhady, A., Khalili, H., Sameti, A.: Development and application of a service design-based process for improvement of human resource management service quality. Bus. Process. Manage J. **27**(2), 459–485 (2021). https://doi.org/10.1108/BPMJ-04-2020-0164

Muhsen, Y.R., Husin, N.A., Zolkepli, M.B., Manshor, N., Al-Hchaimi, A.A.J.: Evaluation of the routing algorithms for NoC-Based MPSoC: A fuzzy multi-criteria decision-making approach. IEEE Access **11**, 102806–102827 (2023). https://doi.org/10.1109/ACCESS.2023.3310246

Nadeem, A., Abedin, B., Cerpa, N., Chew, E.: Digital transformation & digital business strategy in electronic commerce-the role of organizational capabilities. J. Theor. Appl. Electron. Commer. Res. **13**(2), 1–8 (2018)

Nawaz, M.N.I.D.M., Zakai, S.N.Z.S.N., Hassan, M.H.D.M.: An exploration of the factors influences customer response towards banking products and services. Perodicals Soc. Sci. **2**(2), 179–192 (2022)

Norton, D.W., Pine, B.J.: Using the customer journey to road test and refine the business model. Strategy Leadersh. **41**(2), 12–17 (2013)

Pham, T.A.N., Le, H.N., Nguyen, D.T., Pham, T.N.: Customer service co-creation literacy for better service value: evidence from the health-care sector. J. Serv. Mark. **36**(7), 940–951 (2022)

Priyanto, P., Murwaningsari, E., Augustine, Y.: Exploring the relationship between robotic process automation, digital business strategy and competitive advantage in banking industry. J. Syst. Manage Sci. **13**(3), 290–305 (2023)

Purmonen, A., Jaakkola, E., Terho, H.: B2B customer journeys: conceptualization and an integrative framework. Ind. Mark. Manage. **113**, 74–87 (2023)

Rana, J., Jain, R., Santosh, K.C.: Automation and AI-enabled customer journey: a bibliometric analysis. Vision 09722629221149854 (2023)

Rita, P., Eiriz, V., Conde, B.: The role of information for the customer journey in mobile food ordering apps. J. Serv. Mark. **37**(5), 574–591 (2023)

Sadaa, A.M., Ganesan, Y., Yet, C.E., Alkhazaleh, Q., Alnoor, A.: Corporate governance as antecedents and financial distress as a consequence of credit risk. Evidence from Iraqi banks. J. Open Innovation: Technol. Market Complex. **9**(2), 100051 (2023). https://doi.org/10.1016/j.joitmc.2023.100051

Sak, M., Alnoor, A., Valeri, M., Bayram, G.E.: The Role of Digital Transformation on Women Empowerment for Rural Areas: The Case of Turkey. In: Valeri, M. (ed.) Tourism Innovation in the Digital Era: Big Data, AI and Technological Transformation, pp. 91–105. Emerald Publishing Limited (2023). https://doi.org/10.1108/978-1-83797-166-420231006

Schneider, B., Bowen, D.E.: The service organization: human resources management is crucial. Organ. Dyn. **21**(4), 39–52 (1993)

Shen, S., Sotiriadis, M., Zhang, Y.: The influence of smart technologies on customer journey in tourist attractions within the smart tourism management framework. Sustainability **12**(10), 4157 (2020)

Tan, G.W.H., et al.: Metaverse in marketing and logistics: the state of the art and the path forward. Asia Pacific J Mark. Logistics **35**(12), 2932–2946 (2023). https://doi.org/10.1108/APJML-01-2023-0078

Tseng, M.M., Qinhai, Ma., Su, C.-J.: Mapping customers' service experience for operations improvement. Bus. Process. Manag. J. **5**(1), 50–64 (1999). https://doi.org/10.1108/146371 59910249126

Tueanrat, Y., Papagiannidis, S., Alamanos, E.: Going on a journey: a review of the customer journey literature. J. Bus. Res. **125**, 336–353 (2021)

Uhlig, M., Remané, G.: A systematic literature review on digital business strategy. In: 17th International Conference on Wirtschaftsinformatik, Nürnberg, Germany (2022)

Ukko, J., Nasiri, M., Saunila, M., Rantala, T.: Sustainability strategy as a moderator in the relationship between digital business strategy and financial performance. J. Clean. Prod. **236**, 117626 (2019)

Van Vaerenbergh, Y., Varga, D., De Keyser, A., Orsingher, C.: The service recovery journey: conceptualization, integration, and directions for future research. J. Serv. Res. **22**(2), 103–119 (2019)

Wang, Z., Rafait Mahmood, M., Ullah, H., Hanif, I., Abbas, Q., Mohsin, M.: Multidimensional perspective of firms' IT capability between digital business strategy and firms' efficiency: a case of Chinese SMEs. SAGE Open **10**(4), 2158244020970564 (2020)

Wielgos, D.M., Homburg, C., Kuehnl, C.: Digital business capability: its impact on firm and customer performance. J. Acad. Mark. Sci. **49**(4), 762–789 (2021)

Wylde, V., Prakash, E., Hewage, C., Platts, J.: Post-Covid-19 Metaverse Cybersecurity and Data Privacy: Present and Future Challenges. In: Data Protection in a Post-Pandemic Society: Laws, Regulations, Best Practices and Recent Solutions, pp. 1–48. Springer International Publishing, Cham (2023). https://doi.org/10.1007/978-3-031-34006-2_1

Zaidan, A.S., Khaw, K.W., Alnoor, A.: The influence of crisis management, risktaking, and innovation in sustainability practices: empirical evidence from Iraq. Interdisc. J. Inf. Knowl. Manage. **17**, 413–442 (2022)

Zhang, G., Wu, J., Jeon, G., Chen, Y., Wang, Y., Tan, M.: Towards understanding metaverse engagement via social patterns and reward mechanism: a case study of nova empire. IEEE Trans. Comput. Soc. Syst. **10**(5), 2165–2176 (2022)

Zhu, Y., Jin, S.: COVID-19, digital transformation of banks, and operational capabilities of commercial banks. Sustainability **15**(11), 8783 (2023)

The Effect of Financial Readiness on Earnings Retention in Terms of Virtual Technologies: Evidence from UAE Companies

Huda Mohammed Kareem Al-Khafaji[✉] and Hakim Mohsen Mohammed Al-Rubbia

College of Administration and Economic, University of Kufa, Kufa, Iraq
huda.al.khafaji@gmail.com

Abstract. The metaverse is not a concept that can be ignored. The fact that it expands across all industries and businesses shows that it is signaling the next phase of the web. The industry companies are a key player in this area as it is providing entryways into the metaverse through high-quality connectivity and strong network infrastructure. This paper focuses on industrial companies in the UAE in terms of metaverse techniques usages. The current research seeks to determine the impact of financial readiness as an independent variable on earnings retention as a dependent variable. The relationship between the variables was expressed through the hypothetical research model from which the hypothesis was derived. In order to achieve the objectives of the research and verify its hypothesis, it was applied to a sample of the industrial sector companies listed in the Dubai Stock Exchange, which number (7) companies that apply metaverse techniques, through the use of their published financial data for the period (2011–2020). The obtained data were analyzed using Microsoft Excel 2016. The results of the financial and statistical analysis of the industrial sector companies listed in the Dubai Stock Exchange found that financial readiness plays a role in influencing earnings retention based on apply metaverse techniques.

Keywords: Financial Readiness · Earnings Retention · metaverse techniques · Dubai Stock Exchange

1 Introduction

The Metaverse has a similar economic model to the physical world controlled by the economic principles of supply and demand. Goods and services are traded, and transactions are made with the token currency of the platform. With the rise of the digital age, the world around us is changing at a rapid pace. New technology and its trends, like artificial intelligence (AI), machine learning, Internet of Things, virtual reality, and augmented reality, are making a big impact on our lives. As is the norm with new developments, these advancements are taking on new dimensions and tapping into newer, more exciting spaces (Fortune, 2022; Hamid et al., 2022). Due to the intense competition and hostile business environment, the odds of business failures have increased

M. Al-Emran et al. (Eds.): IMDC-IST 2024, LNNS 895, pp. 223–236, 2023.
https://doi.org/10.1007/978-3-031-51716-7_15

dramatically. The term readiness denotes the state through which expected and unexpected future circumstances and events are faced and dealt with. It also means having the skills and good capabilities necessary to deal with future events and learn from them, as well as contribute to what happens (Stevenson, 2010; Eneizan et al., 2019). Pound et al. (2014) believes that financial readiness is one of the most important tools for moving to a better position while achieving effective self-management. It expresses the case in which the successful management of the enterprise supports its financial responsibilities and enhances the personal capabilities of managers and officials to perform their duties with more confidence in the financial resources of the enterprise. In the same context, (Millett et al., 2018; Al-Abrrow et al., 2021) shows how financial readiness consists of a set of distinct capabilities that the establishment possesses, which enable it to provide funds in cash and obtain them at any time, in order to develop any product or service. As financial readiness achieves two main important goals: providing financial flexibility that makes the enterprise able to face various circumstances and help the enterprise increase profits, manage setbacks and avoid exposure to financial difficulties or shocks (Yue et al., 2020; Khaw et al., 2022). Financial readiness is a major strategy for the development of facilities, and a very effective tool in evaluating and allocating budgets. Adoption of the financial readiness process requires careful planning to ensure that failures do not occur (Alsalem et al., 2022).

In addition to the aforementioned, (Hashim et al., 2021; Albahri et al., 2021) believes that financial readiness determines the basic goals of saving for enterprises, strategies for eliminating debts that fall on the enterprise, saving for emergencies, and it is also considered an important strategy that increases the possibility of maintaining reserves (Fadhil et al., 2021; Alnoor et al., 2022). Financial readiness can help mitigate the effect of financial risks and achieve financial stability for industrial companies, through companies adopting financial strategies that allow them to control their financing sources and debt costs, and determine the financial resources required to meet the needs of the company in the future (Shah et al., 2019; AL-Fatlawey et al., 2021). The aim of the study was to investigate strategies for improving the financial readiness of students in Texas to pay for college. Through a coordinated effort between state policymakers, the business and philanthropic communities, students, the public school system, community organizations, and financial institutions. Texas can increase the financial resources available to students with needs and improve the overall financial readiness of students and families to pay for college. As for the study (Ehrlich & Yin, 2022; Gatea and Marina, 2016), the aim of which was to follow up a cross-country comparison of the relative financial readiness of older families in Japan and the Republic of Korea compared to the United States. As a result of what was mentioned above, the main objective of the current study is to know the role of financial readiness and its effect on risk for a sample of industrial metaverse sector companies listed on the Dubai Stock Exchange.

The aim of the research is an extension of some previous studies in the field of defining research variables represented by (financial readiness, earnings retention), as most researchers and practitioners in the field of financial management emphasize that the belief of the senior managements of the research sample companies that apply metaverse techniques in the importance of financial readiness will enable them to achieve their desired goals. Two variables: (financial readiness, earnings retention) were considered, which is evidenced by the absence of any research proving this new trend in the research sample companies that apply metaverse techniques. Also, the study provides a distinguished knowledge contribution in the field of financial management, by defining the research variables represented by (financial readiness, earnings retention). Examining the effect of financial readiness in promoting earnings retention for the research sample metaverse companies. Knowing the impact level of financial readiness in earnings retention.

2 Literature Review

2.1 Metaverse and Economy

Metaverse was primarily based on the convergence of technologies that enable multi-sensory interactions with virtual environments, digital objects and people such as virtual reality (VR) and augmented reality (AR). Hence, Metaverse is an interconnected web of social, networked immersive environments in persistent multiuser platforms. Nowadays, the contemporary iteration of the Metaverse features social, and it provides an immersive experience based on augmented reality technology, creates a mirror image of the real world based on digital twin technology, builds an economic system based on block-chain and tightly integrates the virtual world and the real world (Nesrine & Mohammed, 2023; Manhal et al., 2023). Financial opportunities in the Metaverse have expanded greatly with the advent of Non-Fungible Tokens (NFTs) and virtual assets. NFTs have been crucial in fostering the growth of the digital economy by giving power to producers, facilitating novel forms of ownership, and rewarding active participation on the part of users. However, the long-term viability and inclusion of the Metaverse economy depend on responsible development and awareness of environmental concerns. To fully realize the potential of NFTs and virtual assets within the ever-changing Metaverse, cooperation between stakeholders, creative solutions, and adherence to ethical norms will be required. Entrepreneurs and corporations can find a wealth of untapped potential in the Metaverse. The potential for expansion and innovation is enormous in the realm of virtual real estate, NFT marketplaces, and the incorporation of virtual commerce. Entrepreneurs can use the Metaverse's capacity for immersion and connectivity to build compelling experiences, expand their customer bases internationally, and differentiate their offerings. However, being successful in the Metaverse calls for an in-depth familiarity with user preferences, adherence to responsible and ethical standards, and an openness to new technologies and trends. Entrepreneurs and enterprises may prosper in the rapidly changing digital economy by taking advantage of the opportunities presented by the Metaverse (Aljanabi & Mohammed, 2023; Atiyah and Zaidan, 2022).

Metaverse is an immersive virtual world that simulates and emulates the physical world. People can communicate, collaborate, interact with, and change the elements inside it. Metaverse is often mistakenly seen as a place accessible and interacted with solely by sensory technologies like Virtual Reality (VR) or Augmented Reality (AR), but it is much more than that. Established platforms like PC, gaming consoles and mobile devices also allow access and interaction with the Metaverse. Platforms will become more immersive, and the technology needed to be part of the Metaverse will become cheaper and more affordable. Government intervention may also become another evolution of the Metaverse with the revenue and wealth generated, it will be tempting for governments around the world to get involved and introduce taxes on digital assets (Aldazdi, 2022).

2.2 The Concept of Earnings Retention

The growth of business and technology encourages organizations to compete to create the best products, so that continue to make improvements and obtain sustainable performance. Mistakes in determining the capital structure can put the enterprise in an unfavorable situation, for example, when it uses too much debt, there is a possibility that it will default and lead to bankruptcy (Osesoga & Priska, 2022; Atiyah,2023). The optimal mix of capital structure is to choose the optimal combination of debt and equity that will increase the value of the business, thus maximizing shareholder wealth. However, this choice is not without consequences associated with the overall performance of the enterprise, and also depends on the profit distribution policy of the enterprise (Ugwu et al., 2021). The term earnings retention or what is known as "retained earnings" refers to the net income remaining to the entity after the payment of dividends to shareholders. The entity's retained earnings can be positive (profits) or negative in the case of losses. This retained income can also be kept for reinvestment or debt repayment (Chasan, 2012). Some companies keep more of their profits, so they can reinvest them when they identify viable opportunities, and they can invest mostly in growth companies that have more opportunities as they penetrate the market (Chew et al., 2023; Abbas et al., 2023). However, companies need to conduct appropriate feasibility studies and cost-benefit analysis to avoid misuse of these funds held in unviable investments which can lead to value destruction (Oganda et al., 2022; Atiyah, 2020). Whereas (Oyugi et al., 2019) referred to it as that part of the net profit that is not paid as dividends to the shareholders of the enterprise but is retained for reinvestment purposes through the acquisition of a capital asset or the payment of accrued liabilities such as debts. The accumulated portion of an entity's profits that is not distributed as dividends to shareholders but is instead reserved for reinvestment in the business of the entity (Ball et al., 2020). While it was mentioned (Oketah & Ekweronu, 2020) that it is a financial management tool according to which all profits after taxes are not distributed to the shareholders as dividends, but a part of them is kept or reinvested in the enterprise (Lawal et al., 2022). Thus, we assume that:

H1: There is a statistically significant effect of financial readiness on the variable earnings retention.

H2: There is a statistically significant effect of owned capital on the variable earnings retention.

H3: There is a statistically significant effect of liquidity on the variable earnings retention.

3 Research Methodology

The research hypothesis diagram shows a set of logical relationships between the research variables. The research scheme was designed as a hypothetical scheme, by studying the relationships that were identified through the research problem and its questions.

$$OwnedCapital = TotalAssets - TotalLiabilitie \tag{1}$$

$$Liquidity = \frac{Cash \, and \, an \, near \, Cash}{Current \, Liabilities} \tag{2}$$

While (Lessambo, 2018), investigated the dependent variable for earnings retention as in Eq. (3):

$$RE = RE_0 + NI - D \tag{3}$$

The current research community consists of all industry sector companies that apply metaverse techniques listed on the Dubai Stock Exchange. The sample for the current research consists of (7) companies from the industrial metaverse sector companies listed in the Dubai Stock Exchange, which were chosen because of the availability of data and financial information related to the companies, although choosing the best metaverse development companies was a challenging task, we have come up with great results by considering the following factors:

- Years of experience in metaverse development
- Client testimonials and feedback
- Quality of delivered metaverse products and solutions
- Team size and expertise
- Range of industries served by the company
- The number of successful metaverse projects completed
- Expertise in understanding business requirements and market trends.

Table 1. The Research Sample Companies listed in the Dubai Stock Exchange that Apply Metaverse Techniques.

No	Company Name	Code	Capital/Dirhams
1	Arabia Airlines	AIRARABIA	4666700000
2	Dubai Investment	DIC	4252018000
3	Drake and Scull International	DSI	1070988000
4	Agility Public Warehousing Company	AGLTY	202737000
5	Aramex company	ARMX	1464100000
6	Gulf Navigation Holding Company	GULFNAV	1019209000
7	National Industries Holding Group	NIND	142784000

4 Data Analysis

By observing Table 1, it is clear that the company (AIRARABIA) had an average capital of (5,258,680) dirhams, which is less average than the capital owned by this company for the year (2017), when it reached (6,102,664) dirhams, while the lowest level in the year (2011), when it was (5,249,437) dirhams, and it ranked first among the group. As for the lowest level of owned capital, it was for (Drake & Scull International DSI) in the year (2018), when it was (−4,748,922) dirhams, and the highest level in (2014) was (3,075,986) dirhams, with an average of (263,733) dirhams, which ranked Seventh and final ranking.

By observing Table 2, it is clear that the company (AIRARABIA) had an average liquidity ratio of (0.72), which is an average less than the liquidity ratio of this company for the year (2020), when it reached (0.99), while the lowest ratio was in the year (2012). Where it was (0.06), and it ranked first among the group. As for the lowest level of the liquidity ratio, it was for (Gulf Navigation Holding Company GULFNAV) in the year (2013) when it was (0.03), and the highest level in (2019) was (0.12), with an average of (0.06), which ranked seventh and last.

Table 2. The Results of the Liquidity Index Analysis for UAE Companies.

Years	AIR	DIC	DSI	AGLT	ARM	GULFNA	NIN	Average
2011	0.51	0.08	0.18	0.28	0.62	0.04	0.06	0.25
2012	0.06	0.18	0.21	0.34	0.59	0.05	0.18	0.23
2013	0.81	0.22	0.14	0.37	0.5	0.03	0.12	0.31
2014	0.62	0.48	0.16	0.3	0.78	0.04	0.13	0.36
2015	0.73	0.61	0.09	0.26	0.86	0.05	0.13	0.39
2016	0.78	0.54	0.07	0.21	0.63	0.05	0.09	0.34
2017	0.93	0.36	0.09	0.23	0.53	0.1	0.05	0.33
2018	0.75	0.18	0.03	0.25	0.45	0.08	0.05	0.26
2019	0.97	0.21	0.02	0.27	0.35	0.12	0.04	0.28
2020	0.99	0.22	0.02	0.31	0.45	0.05	0.07	0.3
Average	0.72	0.31	0.1	0.28	0.58	0.06	0.09	0.31

By observing Table 2, it is clear that (Dubai Investments DIC) had average retained earnings of (4,208,680) dirhams, which is an average lower than the retained earnings of this company for the year (2017), when it reached (4,936,167) dirhams, while the lowest level in a year (2012), where it was (3,234,555) dirhams, and it ranked first among the group. As for the lowest level of retained earnings, it was (Drake & Scull International DSI) in the year (2019), when it was (−4,996,454) dirhams, and the highest level in (2014) was (593,554) dirhams, with an average of (−1,458,517) dirhams. Seventh and final ranking. The applied side of this research relied on (7) companies that apply metaverse techniques registered in the Dubai Stock Exchange, the most traded without interruption (2010–2020), which are each of (AIRARABIA, DIC, DSI, AGLTY, ARMX, GULFNAV, NIND) companies. The researchers combined indicator of financial readiness (owned capital and liquidity) and earnings retention indicator.

Table 3. Description of the financial indicators of UAE companies that apply metaverse techniques for the period (2010–2020).

	Owned Capital	Liquidity	Earnings Retention
Minimum Value	281000000	0.02	−4996454
Mediator	2130000000	0.21	428723
Arithmetic Mean	3350000000	0.305	596842
Maximum Value	12300000000	0.99	4936167

The aim of this research is to test a number of hypotheses to show the impact of financial readiness indicators on risk indicators through the linear regression equation, where the best statistical method for estimating the influence coefficients for this model is the least squares method whose estimators are the best unbiased linear estimates if its hypotheses are realized. One of the most important of these hypotheses is that the data is free from outliers. Therefore, the researchers resorted to describing the indicators of the study and revealing whether there are abnormal values. From the results presented in the Table 3 for UAE companies that apply metaverse techniques, we note the following:

1. Owned Capital: Abnormal values are also present in the capital owned by Emirati companies, as is evident in the Fig. 1 and the Table 3 which showed that there is a difference between the median and the arithmetic mean, in addition to that the median

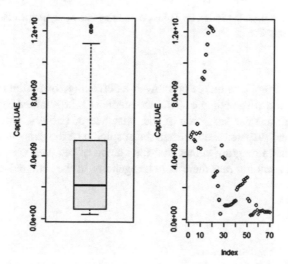

Fig. 1. Outliers in the owned capital index for UAE companies that apply metaverse techniques using the box plot and dispersion.

in the box drawing went to the bottom of the box to indicate that most companies do not It has high ownership capital as it was in 2011 or 2012, but even at the beginning of the period in 2010, and this is very clear in the dispersion chart that Emirati companies started with very high owned capital and then began to decline over time.

2. Liquidity Index: Abnormal values did not appear in the liquidity index data presented in Table 3 as the median is somewhat close to the arithmetic mean. Figure 2 does not show any kind of outliers in this indicator either. However, the liquidity of the UAE companies decreased remarkably in the last period of the study period, as is evident in the dispersion chart and the decrease in the median of the box chart to the downside as well.

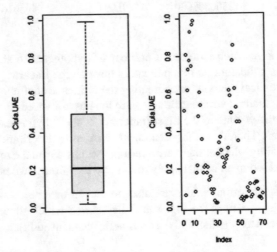

Fig. 2. Outliers in the liquidity index for UAE companies that apply metaverse techniques using the box plot and dispersion.

3. Earnings Retention: This indicator showed a percentage of outliers because the distance between the arithmetic mean and the median of the retained earnings in Table 3 is large enough to raise suspicion of the existence of outliers. However, Fig. 3 the box diagram and dispersion confirmed the suspicion of the existence of such values, and the dispersion diagram shows that the spread of retained earnings was on separate clusters, meaning that there is heterogeneity in the retained earnings of UAE companies.

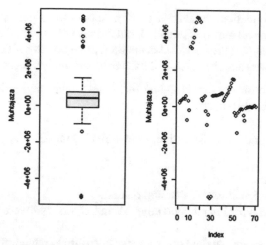

Fig. 3. Outliers in the Earnings Retention index for UAE companies that apply metaverse techniques using the box plot and dispersion.

We conclude from the foregoing that the data set contains abnormal values, and that the scale in Figs. 1 and 3 is completely different from the rest of the figures. Since the values of the capital owned are in millions and the rest of the indicators are only percentages, which makes the process of estimating the impact coefficients of the regression model unrealistic when the indicators are associated or related to each other. Therefore, the researchers resorted to unifying the measurement of all indicators to be distributed in a standard normal distribution according to the following Eq. (4):

$$z = \frac{X - \overline{X}}{\sigma_X} \tag{4}$$

where:

X: financial index.

$\overline{X}$: the arithmetic mean of the financial index.

σ_X: : square root of the variance (standard deviation).

Finally, the variance of the financial indicator with the size of (n) observations is calculated as follows:

$$\sigma_X = \sqrt{\frac{(X - \overline{X})^2}{n - 1}} \tag{5}$$

We rely on the weights of the robust estimation method to multiply the data matrix and then employ it with the weighted least squares method to get rid of the problem of heterogeneity. Equation (5) After the researchers conducted a description of the research data and then consolidated it, we conducted a test of the hypotheses of the UAE companies that apply metaverse techniques, as the researchers studied the test hypotheses by measuring the moral effect of the indicators of financial readiness on the risk indicators each separately and then judging the importance of the model through the value of the coefficient of determination The average until the model was significant. This is because in some cases the model may be significant, but it does not explain the phenomenon in a strong way, and therefore the statistical decision will be positive if the significant effect coefficient is achieved and the value of the determination coefficient is high.

- Let us have the following simple regression model for the relationship between capital owned and retained earnings.

$$Earnings.\text{Retention}_i = \alpha + \beta_1 \text{Capital.IQ}_i + e_i \tag{6}$$

where:

α: slope constant.

β_1: Impact coefficient index of capital owned.

e_i: The random error bound containing all indicators that were not considered by this model.

Abnormal values and random error appear in earnings retention. Equation (6) The emergence of high points of attraction may cause the emergence of another problem called masking and dumping. The mask is that the high point of attraction is effective that it hides under its influence other points of attraction that have not and will not be allowed to be mathematically diagnosed unless the high point of attraction is omitted or its effect is limited. Dumping is a misdiagnosis, meaning when one or more cleans are viewed as poor attractions. Therefore, the researchers used the GM6.IDRGP. RMVN method, which combined the (IDRGP.RMVN) methodology for diagnosis (Uraibi & Al-Husseini, 2021) and the robust GM6 methodology (Uraibi & Haraj, 2022) in one algorithm framework. This methodology allows us to multiply the data of financial indicators with the weights issued by them. These weights would reduce the impact of outliers and points of attraction of all kinds, in a way that does not allow the emergence of the phenomenon of dumping and masking.

As for the estimates of the GM6.IDRGP. RMVN method, we presented in the Table 2 which showed the following:

1. owned capital, which is one of the indicators of financial readiness, has a significant impact on retained earnings, as it explains this phenomenon by (0.577), which is a significant model.
2. The modified coefficient of determination $Adj.R^2$ was (0.985) and the model is significant because the probability value of the F test is less than (0.05).

The following regression model for the relationship between the liquidity index and the earnings retention index Eq. (7). The estimated regression model is written as follows:

$$Earnings.\text{Retention}_i = \alpha + \beta_2 \text{Liquidity.Ratio}_i + e_i \tag{7}$$

We found that the relationship of the correlation coefficient between the liquidity index and earnings retention is inverse and very weak, meaning that any increase in the liquidity index will lead to a decrease in earnings retention, but this decrease will be very small.

The impact coefficient of the liquidity index is significant, its value is (0.43) has a significant impact on retained earnings, as it explains this phenomenon by which is a significant model.

The modified coefficient of determination $Adj.R^2$ was (0.79) and the model is significant because the probability value of the F test is less than (0.05), Statistical inference is the estimated regression model is written as follows:

Let us have the following linear regression model for the relationship between the equity and liquidity indicators and the earnings retention indicator Eq. (8):

$$Earnings.Retention_i = \alpha + \beta_1 Capital.IQ_i + \beta_2 Liquidity.Ratio_i + e_i \qquad (8)$$

Using Eq. (8), the coefficient of the impact of the estimated owned capital amounted to (0.111). This gives us that the owned capital is significant because its probability value (0.000) is less than (0.05). While the estimated impact coefficient of the liquidity index amounted to (0.086), decreased significantly from what it was in the previous findings whose value was (0.43). Therefore, the liquidity index appeared to be insignificant, as its probability value was (0.004), which is less than (0.05). In the sense that indicators of financial readiness have a significant impact on the earnings retention of Iraqi companies to achieve our hypothesis according to the above estimated model, which shows that the impact coefficient of owned capital (0.111) is much than the liquidity impact coefficient with an explanatory power of up to (36%).

5 Conclusions

Financial readiness is an essential component of a company's long-term success and sustainability, in addition, companies that apply metaverse techniques with high financial readiness have greater flexibility in adapting to rapid changes in the economic and market environment. The applied side of this research relied on (7) companies that apply metaverse techniques registered in the Dubai Stock Exchange, the most traded without interruption (2010–2020), which are each of (AIRARABIA, DIC, DSI, AGLTY, ARMX, GULFNAV, NIND) companies. The researchers combined indicators of financial readiness (owned capital and liquidity) and earnings retention indicator. Companies that apply metaverse techniques with high financial readiness have a greater ability to bear risks and face economic and financial challenges, good financial preparation helps companies take advantage of emerging opportunities and expand into new markets. Companies that apply metaverse techniques with good financial readiness can finance investment activities, research and development for innovation, creativity and technological development, good financial preparation enhances companies' ability to achieve sustainable financial returns and increase shareholder value in the long term. Hence, earnings retention is one of the financial strategies used by companies to enhance their ability to expand and grow in the future. Earnings retention may be used as a means of financing research, development and innovation activities in companies that apply metaverse techniques, which contributes to enhancing their competitiveness and creating new products and services. In addition, earnings Retention enhances companies' that apply metaverse techniques ability to deal with emerging financial and economic risks, which achieves greater stability in the company's financial performance. The plan for the Metaverse needs to deal with technical, ethical, and legal issues while making sure that everyone can use it and those users are in control. The open possibilities and new age of connectivity in the Metaverse promise to change the way we connect, create, and work together. By embracing this digital frontier in a responsible way, we can open up a world of endless possibilities, bridge gaps, and create a future that blurs the line between fact and imagination. The Metaverse invites us to go on a journey of discovery and innovation, changing what it means to be connected to other people in this huge, interconnected digital universe. Thus, we suggest that developing the automation tools used in companies to match the development in the world of metaverse and digitization. Also, create the professional financial environment using computer hardware and software, as users will also need to wear devices such as helmets or goggles to interact with the environment and enhancing new skills in the virtual world, such as emotional intelligence and digital skills, for accountants.

References

Abbas, S., et al.: (2023). Antecedents of Trustworthiness of Social Commerce Platforms: a case of rural communities using multi group SEM & MCDM methods. Electronic commerce research and applications, 101322

Ahmed, Aldazdi. (2022). Your personal guide to the metaverse. https://iraqtech.io/your-personal-guide-to-the-metaverse/

Al-Abrrow, H., Fayez, A. S., Abdullah, H., Khaw, K. W., Alnoor, A., Rexhepi, G. (2021). Effect of open-mindedness and humble behavior on innovation: mediator role of learning. Int. J. Emerg. Markets.

Albahri, A.S., et al.: Based on the multi-assessment model: towards a new context of combining the artificial neural network and structural equation modelling: a review. Chaos Solitons Fractals **153**, 111445 (2021)

AL-Fatlawey, M. H., Brias, A. K., Atiyah, A. G. (2021). The role of Strategic Behavior in achievement the Organizational Excellence" Analytical research of the manager's views of Ur State Company at Thi-Qar Governorate". J. Adm. Econ. 10(37).

Aljanabi, M., Mohammed, Sahar Y. (2023). Metaverse: open possibilities. Iraqi J Comput Sci Math., e-ISSN: 2788–7421 p-ISSN: 2958–0544

Alnoor, A., et al.:. (2022). How positive and negative electronic word of mouth (eWOM) affects customers' intention to use social commerce? A dual-stage multi group-SEM and ANN analysis. Int. J. Hum–Comput. Interact. 1–30.

Alsalem, M.A., et al.: Rise of multiattribute decision-making in combating COVID-19: a systematic review of the state-of-the-art literature. Int. J. Intell. Syst. **37**(6), 3514–3624 (2022)

Atiyah, A.G.: The effect of the dimensions of strategic change on organizational performance level. PalArch's J. Archaeol. Egypt/Egyptology **17**(8), 1269–1282 (2020)

Atiyah, A. G. Strategic Network and Psychological Contract Breach: The Mediating Effect of Role Ambiguity.

Atiyah, A. G., Zaidan, R. A. (2022). Barriers to using social commerce. In: artificial neural networks and structural equation modeling: marketing and consumer research applications, pp. 115–130. Singapore: Springer Nature Singapore.

Ball, R., Gerakos, J., Linnainmaa, J.T., Nikolaev, V.: Earnings, retained earnings, And Book-To-Market in the crosssection of expected returns. J. Financ. Econ. **135**(1), 231–254 (2020)

Bănică, E., Vasile, V. (2018). The Export Efficiency Dynamics of Romanian Owned Capital Companies. Limits and Challenges in a Globalised World. The Annals of the University of Oradea, Econ. Sci., 25–38

Chasan, E. (2012). Mid-Size Firms Tap Retained Earnings to Fund Growth. The Wall Street Journal, 3

Chew, X., et al.: (2023). Circular economy of medical waste: novel intelligent medical waste management framework based on extension linear Diophantine fuzzy FDOSM and neural network approach. Environ. Sci. Pollut. Res. 1–27

Ehrlich, I., Yin, Y.: A cross-country comparison of old-age financial readiness in Asian countries versus the United States: the case of japan and the republic of Korea. Asian Dev. Rev. **39**(01), 5–49 (2022)

Eneizan, B., Mohammed, A.G., Alnoor, A., Alabboodi, A.S., Enaizan, O.: Customer acceptance of mobile marketing in Jordan: An extended UTAUT2 model with trust and risk factors. Int. J. Eng. Bus. Manage. **11**, 1847979019889484 (2019)

Fadhil, S.S., Ismail, R., Alnoor, A.: The influence of soft skills on employability: a case study on technology industry sector in Malaysia. Interdiscip. J. Inf. Knowl. Manag. **16**, 255 (2021)

Fortune, (2022). https://www.fortunebusinessinsights.com/blog/top-metaverse-companies-10720

Gatea, A.A., Marina, V.: Higher education funding in Iraq in terms of the experience of particular developed countries. Int. J. Adv. Stud. **6**(1), 8–17 (2016)

Hamid, R.A., et al.: How smart is e-tourism? A systematic review of smart tourism recommendation system applying data management. Comput. Sci. Rev. **39**, 100337 (2021)

Helmcamp, L.: Ready, Steady, Go! Center For Public Policy Priorities, Strategies To Improve Texans' Financial Readiness to Pay for College (2011)

Khaw, K.W., et al.: Modelling and evaluating trust in mobile commerce: a hybrid three stage Fuzzy Delphi, structural equation modeling, and neural network approach. Int. J. Hum-Comput. Interact. **38**(16), 1529–1545 (2022)

Lawal, J.J., Akinrinola, O.O., Olalekan, A.O., Moses, E.C., Gabriel, O.O.: The Impact of retained earnings on the financial growth of pension fund administrator (PFA) companies in nigeria. CJSMS **7**(1), 54–77 (2022)

Lessambo, F. I. (2018). The Sub-Statement of Retained Earnings. In: Financial Statements, pp. 175–180. Palgrave Macmillan, Cham

Manhal, M., Al-khalidi, A., Hamad, Z.: Strategic network: Managerial myopia point of view. Manage. Sci. Lett. **13**(3), 211–218 (2023)

Millett, C.M., Saunders, S.R., Fishtein, D.: Examining how college promise programs promote student academic and financial readiness. ETS Res. Rep. Ser. **2018**(1), 1–24 (2018)

Mansour, N., Abuyousef, M. (2023). Metaverse Technique: Accounting Practice in a Virtual World. J. Res. Financ. Acc., Vol. 08, n.01(/P730–750)

Oganda, A. J., Museve, E., Mogwambo, V. A. (2022). Analysis of retained earnings financing on financial performance of listed manufacturing and allied firms: A Dynamic Panel Approach

Oketah, F. O., Ekweronu, A. C. (2020). Determinants of Retained Earnings of Quoted Manufacturing Firms in Nigeria. Int. J. Manage., Soc. Sci., Peace and Conflict Studies, 3(1)

Osesoga, M.S., Priska, Y.: Factors affecting firms' capital structure: evidence from Indonesia. Educ. Res. (IJMCER) **4**(5), 09–18 (2022)

Oyugi, M.O., Wafula, C., Ngacho, C.: International journals of academics and research. Int. J. Acad. Res. **3**(3), 8–28 (2019)

Pound, C., et al.: 172: The impact of a breastfeeding support intervention on breastfeeding duration in jaundiced infants admitted to a tertiary care centre hospital: a randomized controlled trial. Paediatr. Child Health **19**(6), E94–E94 (2014)

Shah, H.A., Yasir, M., Majid, A., Javed, A.: Impact of networking capability on organizational survival of SMEs: mediating role of strategic renewal. Pakistan J Commer. Soc. Sci. (PJCSS) **13**(3), 559–580 (2019)

Stevenson, A.: Oxford Dictionary of English. Oxford University Press, USA (2010)

Ugwu, P.N.O., Francica, I., Onyekwelu, U.L.: Effect of retained earnings on operational performance indicators of oil and gas firms in nigeria. Adv. J. Bus. Entrepreneurship Dev. **5**(1), 1–9 (2021)

Ukoh, J.E.: Effect of cash holdings on research and development of conglomerates firm in Nigeria. Res. J. of Manage. Pract. **3**(1), 46–57 (2023)

Uraibi, Hassan S., Alhussieny, S. A. (2021). Improvise Group Diagnostic Potential Measures for Multivariate Normal Data. Al-Qadisiyah J. Adm. Econ. Sci. 23(2)

Uraibi, H.S., Haraj, S.A.: Group diagnostic measures of different types of outliers in multiple linear regression model. Malaysian J. Sci. **41**, 23–33 (2022). https://doi.org/10.22452/mjs.sp2 022no1.4

Yue, Z., De Leon, J., Palaoag, T.D.: Assess the level of readiness for adaptation in china: a basis for developing climate-smart agriculture countermeasures. Palarch's J. Archaeol. Egypt/Egyptology **17**(6), 1014–1022 (2020)

Author Index

M. Al-Emran et al. (Eds.): IMDC-IST 2024, LNNS 895, pp. 237–238, 2023.
https://doi.org/10.1007/978-3-031-51716-7

Printed in the United States
by Baker & Taylor Publisher Services